Resource-Oriented Architecture Patterns for Webs of Data

Synthesis Lectures on the Semantic Web: Theory and Technology

Editor

James Hendler, *Rensselaer Polytechnic Institute*
Ying Ding, *Indiana University*

Synthesis Lectures on the Semantic Web: Theory and Application is edited by James Hendler of Rensselaer Polytechnic Institute. Whether you call it the Semantic Web, Linked Data, or Web 3.0, a new generation of Web technologies is offering major advances in the evolution of the World Wide Web. As the first generation of this technology transitions out of the laboratory, new research is exploring how the growing Web of Data will change our world. While topics such as ontology-building and logics remain vital, new areas such as the use of semantics in Web search, the linking and use of open data on the Web, and future applications that will be supported by these technologies are becoming important research areas in their own right. Whether they be scientists, engineers or practitioners, Web users increasingly need to understand not just the new technologies of the Semantic Web, but to understand the principles by which those technologies work, and the best practices for assembling systems that integrate the different languages, resources, and functionalities that will be important in keeping the Web the rapidly expanding, and constantly changing, information space that has changed our lives.

Topics to be included:

- Semantic Web Principles from linked-data to ontology design

- Key Semantic Web technologies and algorithms

- Semantic Search and language technologies

- The Emerging "Web of Data" and its use in industry, government and university applications

- Trust, Social networking and collaboration technologies for the Semantic Web

- The economics of Semantic Web application adoption and use

- Publishing and Science on the Semantic Web

- Semantic Web in health care and life sciences

Resource-Oriented Architecture Patterns for Webs of Data
Brian Sletten
2013

Aaron Swartz's A Programmable Web: An Unfinished Work
Aaron Swartz
2013

Incentive-Centric Semantic Web Application Engineering
Elena Simperl, Roberta Cuel, and Martin Stein
2013

Publishing and Using Cultural Heritage Linked Data on the Semantic Web
Eero Hyvönen
2012

VIVO: A Semantic Approach to Scholarly Networking and Discovery
Katy Börner, Michael Conlon, Jon Corson-Rikert, and Ying Ding
2012

Linked Data: Evolving the Web into a Global Data Space
Tom Heath and Christian Bizer
2011

Resource-Oriented Architecture Patterns for Webs of Data
Brian Sletten

ISBN: 978-3-031-79446-9   paperback
ISBN: 978-3-031-79447-6   ebook

DOI 10.1007/978-3-031-79447-6

A Publication in the Springer series
SYNTHESIS LECTURES ON THE SEMANTIC WEB: THEORY AND TECHNOLOGY

Lecture #6
Series Editors: James Hendler, *Rensselaer Polytechnic Institute*
                Ying Ding, *Indiana University*
Series ISSN
Synthesis Lectures on the Semantic Web: Theory and Technology
Print 2160-4711   Electronic 2160-472X

Resource-Oriented Architecture Patterns for Webs of Data

Brian Sletten
Bosatsu Consulting

SYNTHESIS LECTURES ON THE SEMANTIC WEB: THEORY AND TECHNOLOGY #6

ABSTRACT

The surge of interest in the REpresentational State Transfer (REST) architectural style, the Semantic Web, and Linked Data has resulted in the development of innovative, flexible, and powerful systems that embrace one or more of these compatible technologies. However, most developers, architects, Information Technology managers, and platform owners have only been exposed to the basics of resource-oriented architectures. This book is an attempt to catalog and elucidate several reusable solutions that have been seen in the wild in the now increasingly familiar "patterns book" style. These are not turn key implementations, but rather, useful strategies for solving certain problems in the development of modern, resource-oriented systems, both on the public Web and within an organization's firewalls.

KEYWORDS

REST, Web, Semantic Web, services, RDF, SPARQL, Linked Data, patterns

With all my love to Kristin and Loki.

SWEETEST love, I do not go,
For weariness of thee,
Nor in hope the world can show
A fitter love for me

John Donne (1572–1631)

Contents

List of Figures

Preface

It is hard to believe that we are nearing the 20th anniversary of the fabled "Gang of Four" *Design Patterns* book. Enthusiasm for the "Pattern" book style has waxed and waned in the interim. While the style has been done and overdone, it is still a comfortable approach to distilling bite-sized nuggets of design discussion. Every entry is given a name, a motivation, examples, and a discussion about potential positive and negative outcomes. The structure of the narrative reminds us that choices have consequences and designs have intended uses. It encourages us to be thoughtful and cautious in the techniques we employ to solve real-world problems. It also gives us a high-bandwidth way to describe a significant design element of a system by name.

Soon after discovering the elegance of patterns, many developers fall into the "Everything is a Pattern" Anti-Pattern. They misinterpret patterns as components and consider them building blocks. They believe that patterns give them a pass on actually thinking about the systems they are building. They seek frameworks that implement the patterns for them, avoiding even more work.

That is not the goal of this book. The resource abstraction is about shielding clients from unnecessary details. Resource-oriented patterns are not about shielding developers from thinking about what they are building. They represent Uniform Interfaces, but not necessarily Universal Interfaces. Their applicability may vary widely-depending upon the context of your requirements.

It is also hard to believe that we have just recently passed the 20th anniversary of the World Wide Web. What would have at one time seemed inconceivable to nearly everyone has become commonplace to most.

I am convinced that we are not done learning from the deliberate and cautious choices made at the beginning of this journey. The migration from static documents to dynamic applications certainly involved new technologies (e.g., AJAX, JavaScript libraries, Cascading Stylesheets, etc.), but it was enabled by the underlying pieces of logically named *Uniform Resource Locators* (URLs), the *Hypertext Transfer Protocol* (HTTP) and *Hypertext Markup Language* (HTML). Even as we consider modernizing protocols such as SPDY and HTTP 2.0, we are doing so with the goal of keeping the infrastructure consistent.

It is to the immense credit of Tim Berners-Lee, and to those who helped him to design and build The Web, that this fundamental communication pattern was maintained. When you want a document on the Web, you simply ask for it by name. Finding the name is a separate issue that we will revisit later, but the notions of disambiguation and resolvability are wrapped up in the design of *Uniform Resource Locators*.[1] The name of the thing is not only its identifier in a global context, it is also the handle by which you request it. Name. Point. Request.

---

[1] http://tools.ietf.org/html/rfc3986

The scope of what we can address, of course, includes documents. But we refer to *resource* locators, not simply document locators, because we have more varied types of information to address. These can also include data, services and concepts. Not all resources are directly addressable, but we have a common scheme for addressing all of the different things we care about.

Another trick the Web pioneers pulled in their design was to give people the ability to choose the names for the things they published. You did not need to seek the permission of a central authority to name something and share it. It is remarkable (and largely unexpected) that giving people this ability did not devolve into chaos. Entire industries had previously emerged to manage the centralized control of identifiers such as International Standard Book Numbers (ISBNs), International Standard Audiovisual Numbers (ISANs), and Digital Object Identifiers (DOIs). The benefit these organizations promised was guaranteed uniqueness. But on the Web, anyone can give a name to a document in a domain they control without permission and not have to worry about conflicts.

The combined decisions of allowing people to name their own resources, publish their content at will, and letting those names be used in the resolution process were crucial to the success of the Web. Sharing documents became as easy as sharing the name of the document. This ultimately changed our whole notion of how to share information in a networked environment. We began to talk more about information resources and their identity and less about document formats and storage. This level of indirection opened up paths to dynamic content generation, technology migration strategies, context-specific responses to individual clients, and more. The Web of Documents has become the Web of Data.

These technologies combined seamlessly to bring us a platform usable by scientists, knowledge workers, grandparents, and children. It was not an accident that this happened. Tim Berners-Lee, Roy Fielding, Dan Connolly, Dave Raggett, and others from the *Internet Engineering Task Force* (IETF) and *World Wide Web Consortium* (W3C) toiled to create a platform unencumbered by patents and displaying the architectural properties of loose-coupling, scalability, flexibility, and extensibility.

And yet, it still requires a surprising amount of motivation to get people to think about adopting these technologies within firewalls as well. We forget that Tim's original proposal was written in the context of solving the information sharing needs of a single (albeit complex) organization, the European Organization for Nuclear Research (CERN).

In his storied proposal,[2] he said:

> "If a CERN experiment were a static once-only development, all the information could be written in a big book." Berners-Lee [1989]

Clearly, this was not the case. In trying to do fundamental science, new participants, new strategies, new results, and new experiments needed to be constantly accepted by the system. Given that CERN brought researchers from around the world and interacted with other academic, commercial, and governmental institutions, no social or organizational policies could force standardization

---

[2]Deemed "vague but exciting" by his supervisor. `http://info.cern.ch/Proposal.html`

among all the participants. A technological solution that embraced change was proposed. After the proposal got no traction for quite some time, Tim was eventually given permission to work on it on the side. The rest is history.

Most of the world embraced the vision whole-heartedly because it allowed them to participate regardless of the technology choices they had made. Different operating systems and platforms were melded into an interoperable environment where anyone could contribute. Broken links were tolerated and localized to specific interactions. The decentralized approach of logically-connected resources allowed demand to spike unpredictably.

All of this is to say that the approach looked at the world of information integration differently. Change was expected. Breakage was expected. Consensus was not required. Central authority could be embraced or not.

In the years since, we have seen a spate of activity around the development of resource-oriented architectures. Most have gotten it wrong because they bring with them the baggage of code-focused world views of excessive coupling. The ones who succeeded did so by aligning themselves with the vision of The Web.

It may be premature to catalog resource-oriented patterns, but these are solutions I have seen in the wild. The goal is to help others make deliberate and cautious choices in their own resource-oriented systems so that they, too may align themselves with an architectural world view that has been so successful.

In Chapter 1, we introduce four patterns that act as primary sources. These represent how we choose to organize our information in ways that embrace change. In Chapter 2, we move to the secondary structures of resources being applied to other resources. In the process, we consider how we can transform and manipulate content available to our creativity but outside of our control. Finally, in Chapter 3, we end with four patterns that connect the resource view of the world to the realities of modern business systems.

It is difficult to describe a dynamic and flexible ecosystem solely in the linear narrative of prose, so I have developed running examples of each of the patterns in two modern, resource-oriented environments. As the world's only fully resource-oriented software environment, NetKernel (`http://netkernel.org`) is an obvious choice to explore these ideas. Additionally, the Restlet (`http://restlet.org`) environment elevates Web abstractions in a more conventional Java-based framework.

You will find these examples and instructions for getting them up and running here: `https://github.com/bsletten/roa-book-examples`. You are encouraged to experiment with the sample implementations to deepen your insight into how they can help you make deliberate choices in building your own resource-oriented architectures.

Don't forget: Thinking is still required.

I would like to thank the series editors Jim Hendler and Ying Ding for their encouragement and stewardship throughout the process of writing this book. I would also like to thank Mike Morgan of Morgan & Claypool for his patience and the opportunity to contribute to the growing body of work in this series.

The ideas in this book would never have germinated without the inspiration provided by Peter Rodgers and Tony Butterfield of 1060 Research Ltd. (http://www.1060research.com)

I would not have gotten the practice of talking about these patterns without the opportunity afforded me by Jay Zimmerman and the NoFluffJustStuff (http://nofluffjuststuff.com) conference series. Being given the chance to have a conversation with committed developers 20-30 times per year is a gift. The friendship, support, and inspiration of my fellow "FluffTalkers" past and present is a part of that gift.

Finally, nothing is worth anything without the love and support of my wife, Kristin, and Loki, the God of Mischief. I am frequently too busy working on my things and travel too much, but I derive no greater joy than returning to spend time with the two of you.

Brian Sletten
Bosatsu Consulting, Inc.
April 2013

CHAPTER 1

Informational Patterns

1.1 INTRODUCTION

The Informational Patterns identified here are loosely categorized as those that serve information-sharing and organizational purposes. The *Information Resource* Pattern itself really needs no motivation. It has an established reference implementation called the *World Wide Web*. The logically named, interlinked HTML documents became the way of sharing information with collaborators, partners, clients, and the general public. As we grew discontented with static documents, we found ways to make them dynamic and to reflect the contents of our data stores in user-friendly and navigable ways. This in turn fueled a hunger for zero-installation, dynamic systems. Web sites became web applications.

From there, we started to consider longer-term and more flexible ways of organizing our information. The *REpresentational State Transfer* (REST) architectural style got us to think about expanding our world view of resources to include non-document sources. Resource identifiers were applied to arbitrary partitions of information spaces. We needed to think about how to organize collections of information resources in meaningful and reusable ways. We eventually pursued full-blown Webs of Data; interlinked and interoperable. We needed to name not only our entities, but also the relationships that connected them. As consumers of these resources, we wanted a greater say in defining what was important to us. We wanted to be able to share what we valued with those we valued, with the same ease as any other location on the Web.

The shifts in thinking that brought us these technologies were crucial but subtle. They were so subtle that we failed to notice that the *World Wide* aspects of the *Web* did not preclude it from being equally applicable in smaller environments such as the Enterprise. After a decade or more of service-only thinking, we were ready to adopt the resource abstractions as a superset of the documents, data, services, and concepts that comprised our business.

1.2 INFORMATION RESOURCE

1.2.1 INTENT

The *Information Resource* Pattern is the basic building block of the Web. It provides an addressable, resolvable, relatively course-grained piece of information. It should generally have a stable identifier associated with it and optionally support content negotiation, updates, and deletions.

Figure 1.1: Basic Information Resource Pattern.

1.2.2 MOTIVATION

The main feature of the *Information Resource* Pattern is the use of logical names to identify them. These names serve the dual purpose of identification and manipulation. We are using the name as an identifier in global context as well as name as a handle with which to interact with the information resource. The logical identifier isolates us from implementation details. It keeps us from thinking about code or even services. We are talking about the information itself. This approach elevates information to its rightful place as the leading component of the systems we build. We give identity to the things we care about and we ask for it through these names. We do not need to write a query or invoke a separate service to retrieve the thing we seek.

On the public Web, we recognize the ability to refer to a specific document:
`http://bosatsu.net/index.html`.

We can think beyond documents, however. Whatever domain we work in, we can imagine translating the data elements we care about into named resources:

`http://example.com/account/user/jsmith`,
`http://example.com/employee/id/12345`,
`http://example.com/order/id/101-276-32`, or
`http://example.com/product/id/upc/800314900528`.

These names represent good, long-lived, stable identifiers for these disparate domains. A reference to one of these resources today could still be valid a year from now—or ten years from now. The reason for this is because we also lose any coupling to how the information is produced. We do not know what, or even who, is producing the result. Honestly, we do not care. As long as something is still responding to the requests, we as clients are in good shape.

The separation of the name of the thing from the production process is only part of the story, however. The shape of the information is also free to change based upon the context of the request. Developers, business analysts, accounting departments, and end users may all lay claim to a preferred form for the information. This is as it should be. A development-oriented format like XML or JSON is useless to someone who wants to look at it in a spreadsheet or in a browser.

The separation of the name from the form does not require a new identity. We can turn the request into a negotiated conversation and, where possible, satisfy the needs of disparate consumers from a single location. Avoiding data *extraction, transformation, and loading* (ETL)[1] steps reduces the burden of having multiple copies of our information repositories. We can achieve the goals of a *Master Data Management*[2] solution while simultaneously avoiding the unnecessary limitations of a prescribed format.

If we decide to update the information, we can send a new copy back to the location from where we got it. Again, no separate service or software is required on our part as long as the request is handled. The general mechanism of pushing and pulling on information resources is generalizable and somewhat universally applicable. We quickly recognize that, as consumers or producers of information, we do not need to come up with separate ways of handling the information types we care about. It can all basically work the same way. Just like the Web does.

We ignore many details with this generalization, but we will address them separately in other sections. The main benefits we get from this pattern are loose-coupling, scalability, and predictability.

1.2.3   EXAMPLE

The Web itself is the single best example of the *Information Resource* Pattern we have. We generally understand how and why it grows and evolves over time, but we tend to sell the pattern short. People draw distinctions between the public Web and what happens behind firewalls. They see artificial boundaries between sharing documents and sharing data.

Architectural styles such as the *REpresentational State Transfer (REST)* broaden the applicability of the pattern by defining strategies for sharing information in scalable, flexible, loosely coupled ways. We use stable, HTTP identifiers to define the resources. We manipulate them via HTTP verbs. We extend the architecture so that it supports new shapes for the information with content negotiation.

In the REST architectural style, the body of what is returned for an information resource should be a self-describing, hypermedia message that yields the *affordances* available to the client. These include links for the resource in alternate forms, the discovery of links to manipulate the resource further, or means to find related resources. The abstraction remains the information resource, however. This can be a pure abstraction over the domains reflected to minimize the visibility by the client into actual production considerations.

A common problem when developers first attempt to produce REST APIs is to think of an information resource as equivalent to a domain object. They employ simple conversions tools to emit XML or JSON serializations of the object structure for the representation. Perhaps it is non-intuitive, but what is easy for the developer is often bad for the API. The resource is not a domain object. By that, we mean that the representation of the resource should not be tightly bound to an existing domain model. The reason for this is that we want stability and predictability in our APIs. Changes to the domain model will automatically flow through to changes in the serialization. When that acts as your API, your clients will break. This is the opposite of what we want. Instead,

---

[1]http://en.wikipedia.org/wiki/Extract,_transform,_load
[2]http://en.wikipedia.org/wiki/Master_data_management

we need to pick a form that *represents* the information resource without being tightly coupled to how it is stored or produced. This is the "R" in REST's *REpresentation State Transfer*.

A first attempt at this might look something like the following.[3] To retrieve an account from `http://example.com/account/id/12345`, the client would issue an HTTP GET request to that URL:

```
1  GET /account/id/12345
2  Host: example.com
```

and the server might respond with a language and platform-independent representation of an account:

```
1   HTTP/1.1  200  Ok
2   Content-Length: 519
3   Expires: Wed, 27 Mar 2013 12:15:00 GMT
4   Date: Wed, 27 Mar 2013 12:14:00 GMT
5   Content-Type: application/xml
6   Server: NetKernel [1060-NetKernel-SE 5.1.1] - Powered by Jetty
7   Cache-Control: max-age=600
8   Last-Modified: Wed, 27 Mar 2013 12:13:00 GMT
9   Etag: "221207d"
10
11  <account id="12345">
12     <name>bsletten</name>
13     <status>gold</status>
14     <recentOrders>
15        <order id="141234541234">
16           <items>...</items>
17        </order>
18        <order id="452354234534">
19           <items>...</items>
20        </order>
21     </recentOrders>
22  </account>
```

With an advertised MIME type of application/xml, there is no reason for a client to know that this is an account, or what to do with it. Certainly a programmer could write the client to know to treat it like an account. It expects an association between the URL `http://example.com/account/id/12345` and an XML representation of the account. Again, this seems like a minor thing but it absolutely flies in the face of how the Web works. When you point your browser at any random website, it does not "know" anything. It issues a standard, semantically constrained verb, called GET, to a named element and responds to what comes back. "Oh, an HTML document! I know how to parse that." or "Oh, a plain text file, I know how to display that." It is reactive. It does not know. Knowledge is a form of coupling. If the server changes what it returns, the client breaks. The Web would never have worked like that.

Now, there is not much a browser can do with a "text/plain" file but display it. An HTML document on the other hand is filled with *affordances*. This is a design term for something that

---

[3]We are ignoring any kind of security model at this point.

allows you to achieve a goal. Discovering an image allows you to see it in the current document. Discovering a link allows you to visit another document. Everything that the Web allows us to do is a function of the affordances presented to us (e.g., login, submit queries, buy stuff, check headlines, play games, chat). These are not presented in domain-specific ways. They are presented as options for us to choose in the body of a standard representation.

Strict REST conformance requires representations to be self-describing hypermedia messages. The client does not know what it is going to get, it reacts to what is returned. It knows how to parse standard types. And from there, it discovers what options to present to the user (if there is one).[4] While there are definitely reusable hypermedia formats (HTML, Atom, SVG, VoiceXML), there are currently no standards-based, general-purpose hypermedia formats[5] for representing arbitrary informational content. Some Restafarians advocate the use of HTML for even non-presentation purposes, but that is not a widely supported position.

To counter this current gap, many pragmatic members of the community advocate the creation of custom MIME types. While it is not ideal, it is a way to document a representation format. Rather than identifying your account representation as application/xml, you might define a new format and call it application/vnd-example-account+XML. This format is simply a convention. The application/vnd part of the name indicates that it is for a specific custom use. The example segment identifies your organization (e.g., we are assuming a domain name of example.com for most of our examples). The account portion identifies the resource within your domain and the +XML gives you notice that it is an XML serialization. The same custom type in JSON might be application/vnd-example-account+JSON. You are encouraged to register your custom names with the Internet Assigned Numbers Authority (IANA),[6] but there is nothing to force you to do so.

The media type should be designed to support hypermedia links. This should identify the resource itself, as well as its related resources. Clients will be able to "follow their noses" through your hypermedia representations. In the following example, we can find our way back to the account resource itself (useful if we did not fetch it initially but were given the representation as part of an orchestration), its recent orders, as well as individual orders.

```
1   <account id="12345">
2       <link rel="self" href="http://example.com/account/id/12345"/>
3       <name>bsletten</name>
4       <status>gold</status>
5       <recentOrders href="http://example.com/order/account/id/12345/recent">
6           <order id="141234541234" href="http://example.com/order/id/141234541234">
7               <items>...</items>
8           </order>
9           <order id="452354234534" href="http://example.com/order/id/452354234534">
10              <items>...</items>
```

[4]Many people think that hypermedia affordances are only useful when there is a human involved. That is simply not true. These links are discoverable by software agents and bots as well.

[5]Although there are some proposals on the table, most notably HAL (http://stateless.co/hal_specification.html) and SIREN (https://github.com/kevinswiber/siren), we will revisit these formats in the discussion of the *Workflow Pattern* (3.5).

[6]http://www.iana.org/assignments/media-types

```
11          </order>
12        </recentOrders>
13    </account>
```

Now, when you return:

```
1    Content-Type: application/vnd-example-account+XML
```

your client knows how to handle this type, assuming you have added the capability to your application. In this sense, your client applications become more like browsers. They react to what they are given. They know how to handle certain types of representations. And, they find everything they need in the bodies of what is returned. There is no out-of-band communication being used to tell it what to do. For a larger discussion on hypermedia and representation design, please see Amundsen [2011] and Richardson [2013].

The HTTP verbs give us the ability to interact with resources directly. As we have seen, we can retrieve a resource with the GET verb. We update resources with POST, PUT and PATCH, depending upon the nature of the update. We can remove resources by issuing DELETE requests. We can discover metadata about a resource with the HEAD method. We can discover what we are allowed to do to a resource with the OPTIONS method. The verbs are classified based upon whether the requests associated with them can be repeated. As a non-destructive, safe, read-only request, there should be no problem issuing multiple GETs. This is important because, in the face of potential temporary network failure, we do not want the HTTP client to give up and create an application level error. If it is able to retry the request, it should. This property is called *idempotency* and it is a key reason Restafarians get mad when you violate this principle.

A GET request that actually changes state (a violation committed by way too many developers who do not know or care what they are doing) messes up the potential for caching results and can have unexpected results if the request is bookmarked.[7]

PUTs and DELETEs are intended to be idempotent. PATCH can be made idempotent with the use of Etags and conditional processing. POST, as a mostly semantically meaningless method, is not. It is extremely useful for the resilience of your system to understand and correctly leverage idempotency where you can.

The response codes give us an indication of success or failure. "200 OK" means the request is successful. "204 No Content" is used to indicate that the request was successful but there is no other response. A "401 Unauthorized" means, without more information, you are not allowed to see the resource. A "403 Forbidden" means you are simply not allowed to see it.

The main thing to remember is that none of these identifiers, verbs, or response codes are specific to a particular domain or application. The Web architecture has a *Uniform Interface*. Your browser is unaware of whether you are checking headlines or chatting on a forum. It all works the same way. Our goal in using the *Information Resource* Pattern is to make the various aspects of our

---

[7]It can also open you up to a form of attack called *Cross Site Request Forgery* (CSRF). See http://en.wikipedia.org/wiki/Cross-site_request_forgery for more information.

systems also work the same way. As we have discussed above, particular types may offer different capabilities or affordances, but they all generally function alike.

Finally, not everyone will want an XML serialization of the accounts, orders, customers, or whatever else we are talking about. We also want to support content negotiation at the resource level to allow people to ask for information in the shape that they want it. This alternate form will still be served from the same location. We do not need to break existing clients to support new formats. We do not need to support all of the formats at once. We can incrementally add capabilities as they are requested, without having to export the data and serve it up from somewhere else in another form.

The HTTP GET request allows a client to specify a desired representation type:

```
1  GET /account/id/12345
2  Host: example.com
3  Accept: application/vnd-example-account+JSON;q=0.9,
           application/vnd-example-account+XML;q=0.5
```

Here the client is expressing a preference for the JSON form of our accounts, but indicates that it will also accept the XML form if we are unable to satisfy its first choice. The most important concept here is that the same name, http://example.com/account/id/12345, can be represented in multiple ways. Unfortunately, HTML does not support the *Read Control* (CR) HFactor,[8] which means it cannot use content negotiation natively. Instead, it would have to rely on a JavaScript invocation of XMLHttpRequest:[9]

```
1  httpRequest.setRequestHeader('Accept', 'application/vnd-example-account+JSON');
```

Traditionally, developers use suffixes to trigger this behavior. For the XML form, they might publish http://example.com/account/id/12345.xml. For JSON, http://example.com/account/id/12345.json. There is fundamentally nothing wrong with this approach, but it can make it more difficult for systems that your client interacts with (i.e., as part of an orchestration across resources) to negotiate the resource into a different representation. The other issue here at a larger resource-oriented level is that you are forking the identity of the resource. Every new suffix stands for a new resource. Any metadata captured about this resource with something like the *Resource Description Framework* (RDF) would need to be linked to all of the forms. There are ways of managing this, we just end up working harder than we need to. If possible, you should consider using the purely logical identifier not bound to a particular type, and content negotiate as needed.

We are alerted to one final problem with this approach by Allamaraju [2010]. If we are going to vary the response type for the same identifier based upon Accept headers, we need to indicate that in our response. We need to include a Vary response header to indicate that intermediate proxies should identify resources by compound keys involving both the name and the MIME type. Otherwise, they might accidentally return the wrong type to a client, thinking they are asking for what has already been cached because they use the same name.

---

[8] See Amundsen [2011] for the definition of the various *Hypermedia Factors* (HFactors). We will discuss them more in the *Workflow Pattern* (3.5).

[9] This is also how you have to specify non-idempotent methods like DELETE and PUT when using HTML, because it lacks support for the *Idempotent Links* (LI) HFactor.

1.2.4   CONSEQUENCES

Any abstraction has a cost. The choice of adopting an *Information Resource* Pattern may have implications involving performance, network traffic, and security. TCP/IP is a connection-oriented protocol that is expensive to establish in the face of high-speed requests and transactions. Advances in protocols such as SPDY[10] and HTTP 2.0[11] are attempting to mediate these issues by reusing the TCP/IP connection for long periods in order to amortize the set-up cost over more interactions.

Beyond the performance issue, the purity of the abstraction may hide the complexity of interactions between multiple backend systems. State change notifications and transactional boundaries may not fit as cleanly into the abstraction. This pattern requires stateless interaction, which may be an issue depending on the nature of the clients that are using it.

From the client's perspective, the *Information Resource* Pattern pushes a significant amount of complexity to it. The server is free to support various representation forms (or not), so the client may have to do the integration between resources to merge multiple XML schemas, JSON, and other data.

See also: *Collection Resource* (1.3), *Guard* (2.2), *Gateway* (3.2).

1.3   COLLECTION RESOURCE

Figure 1.2: Collection Resource Pattern.

1.3.1   INTENT

The *Collection Resource* Pattern is a form of the *Information Resource* Pattern (1.2) that allows us to refer to sets of resources based upon arbitrary partitions of the information spaces we care about. To keep client resources from having to care about server-specific details, these resources should provide the mechanism for relationship discovery and pagination across the collections.

---

[10]http://en.wikipedia.org/wiki/SPDY
[11]http://en.wikipedia.org/wiki/HTTP_2.0

1.3.2 MOTIVATION

In Section 1.2, we discovered the benefit of adopting the *Information Resource* Pattern for specific pieces of information. Now, we would like to expand our expressive capacity to talk about collections of these resources. A useful way to think about these is as sets and members of sets. We would like to be able to give names to whatever sets make sense to our domain. Consider a commerce application that deals with accounts, products, promotions, etc. Each one of these general categories can be thought of as the set of all entities of that type. Because we are dealing with logically named resources, we can imagine defining:

http://example.com/account,

http://example.com/product, and

http://example.com/promotion.

The choice of whether to use plural terms to refer to the sets is likely to be subjective, but we will use the singular form, and understand that we are identifying a potentially infinite set of resources. We are not yet worrying about the practicality of such large sets, we are simply creating identifiers.

Within an individual space, we might like to partition it along arbitrary boundaries. We might like to refer to the set of all new accounts, the best accounts, the overdue accounts, etc. As we are breaking them up on a status partition, it makes sense to introduce that notion into our identifiers:

http://example.com/account/status/new,

http://example.com/account/status/top, and

http://example.com/account/status/overdue.

These partitions make sense, as does giving them identifiers that behave like our singular information resources. These identifiers will map to a REST service endpoint that will translate the request into a backend query. As consumers of this information, we do not care what kind of data store is being queried nor how it happens. All we care about is giving meaningful business concepts stable identifiers and being able to resolve them.

The resolution process works like any resource request. An HTTP GET request to the identifier returns a representation of the collection resource. The body of the response is mostly about providing linkage to the individual resources, however, it is important to think about how a client might use the resource. If the resource handler returned something such as the following:

```
1  <accounts href="http://example.com/account/status/top">
2    <account href="http://example.com/account/id/12345"/>
3    <account href="http://example.com/account/id/34246"/>
4    <account href="http://example.com/account/id/77323"/>
5  </accounts>
```

then there is very little the client can do without resolving each of the individual resources. From a performance perspective, we are inducing an excessive number of roundtrips before anything can be displayed. You will learn to strike a balance between sending too much information in a representation payload and not sending enough.

A slightly more verbose representation:

```
1   <accounts href="http://example.com/account/status/top">
2     <account id="12345" username="jojalehto" status="active"
3         href="http://example.com/account/id/12345"/>
4     <account id="34246" username="bkemp" status="active"
5         href="http://example.com/account/id/34246"/>
6     <account id="77323" username="bluu" status="disabled"
7         href="http://example.com/account/id/77323"/>
8   </accounts>
```

gives a client the ability to present some amount of information about the members in this particular collection without requiring a trip back to the server. The user might be presented with a sorted list, color-encoded by status, or whatever made sense.

1.3.3   EXAMPLE

While it might be tempting to build a custom MIME type for each resource collection we have, there are definitely existing choices we can use. Unless there is a specialized need that is not met by existing formats, it is a good idea to reuse established types. We could use the *Atom Publishing Protocol*,[12] which has support for collections of resources. We could also use the *Open Data Protocol* (OData),[13] which uses either JSON or Atom to create uniform APIs. Either of those would be good choices, but we will try a registered type that is slightly lighter weight.

Mike Amundsen[14] designed the application/vnd.collection+json[15] format to have a reusable collection type expressed in JSON. He has registered it with the IANA,[16] so it is a legitimate and proper hypermedia format supporting six of his nine HFactors.[17] In addition to supporting lists of resources, it also supports arbitrary key/value pairs for the resources, related links, query templates, and a discoverable mechanism for creating new item elements in the collection.

Here we reimagine our collection of top accounts as a application/vnd.collection+json representation:

```
1   {
2       "collection" :
3       {
4           "version" : "1.0",
5           "href" : "http://example.com/account/status/top",
6
7           "items" : [
8           {
9               "href" : "http://example.com/account/id/12345",
10              "data" : [
11                  { "name" : "id", "value" : "12345" },
12                  { "name" : "username", "value" : "jojalehto" },
13                  { "name" : "status", "value" : "active" }
```

---

[12]http://tools.ietf.org/html/rfc5023
[13]http://www.odata.org
[14]http://amundsen.com/blog/
[15]http://amundsen.com/media-types/collection/
[16]http://www.iana.org/assignments/media-types/application/vnd.collection+json
[17]Again, see Amundsen [2011] for a discussion of Hypermedia Factors (HFactors).

```
14          ],
15          "links" : []
16       },
17       {
18          "href" : "http://example.com/account/id/34246",
19          "data" : [
20             { "name" : "id", "value" : "34246" },
21             { "name" : "username", "value" : "bkemp" },
22             { "name" : "status", "value" : "active" }
23          ],
24          "links" : []
25       },
26       {
27          "href" : "http://example.com/account/id/77323",
28          "data" : [
29             { "name" : "id", "value" : "77323" },
30             { "name" : "username", "value" : "bluu" },
31             { "name" : "status", "value" : "disabled" }
32          ],
33          "links" : []
34       }
35    ],
36
37    "links" : [
38       { "href" : "http://example.com/account/status/definitions", "rel" :
            "definition" }
39    ],
40
41    "queries" : [
42       {
43          "href" : "http://example.com/account/top/search",
44          "rel" : "search",
45          "prompt" : "Search top accounts",
46          "data" : [
47             { "name" : "username", "value" : "" },
48             { "name" : "status", "value" : "" }
49          ]
50       }
51    ],
52
53    "template" : {
54    }
55  }
56 }
```

The URL for the collection itself is identified by the outer href key. From there we have the individual items of the collection. Note that we do not explicitly identify a type for the items, but that might not be an unreasonable extension. If there were related links for each item, they could be specified individually. We also have a related link for the collection. It points to a description of the various account classifications. It is here that we could imagine support for pagination over very large collections:

```
1  "links" : [
2    { "href" : "http://example.com/account/status/definitions", "rel" :
         "definition" },
3    { "href" : "http://example.com/account/status/top;page=2", "rel" : "next" }
4  ],
```

The clients will not have to know how to paginate across arbitrary collections, they will simply discover links related to the collection with a rel type of next or previous. The server still drives URL layout, which is what we want in a hypermedia system.

We do not have values for the template key because we do not explicitly add items to the top collection. They are implicitly added based on attributes on the account. If this were another type of collection, however, in the template section we could discover links, required key value pairs, and a prompt for collecting information from the user. The collection also can have an optional error key that lists related error conditions.

Hopefully it is no great leap to imagine different collections of resources using the same format. These could either be other collections of different resource types, or other partitions of the same resource type (e.g., overdue accounts, new accounts, etc.). The benefit of a reusable type is, obviously, that you only have to write the code to support it once. The format is extensible because we can add arbitrary key/value information for the collection or each individual item. Clients can discover the URLs in the body of the representation so that they remain uncoupled to a particular server layout.

1.3.4   CONSEQUENCES

There are not many substantial consequences, positive or negative, from the use of the *Collection Resource* Pattern for information resources. The ultimate trick is whether the level of detail returned per item is sufficient for the clients' needs or not. It is important to strike an appropriate balance for the general case, but this may leave specific clients with either too much or too little data. Too much data is a waste of bandwidth, or perhaps an increased parsing load for large collections. Too little data is probably the more serious issue. As we identified with accounts in Section 1.3.2, if we just returned links for each account, any meaningful use of the data will require expensive network GET requests for each element. That is still true when using a standard format such as application/vnd.collection+json. If there are required attributes on each item that are not conveyed through the key/value capabilities, a client will have to fetch the individual resources to get what it needs. That would most likely be prohibitively expensive and generally represent a mismatch in representation. It may be worth establishing a separate URL endpoint with a more detailed view of the item data to solve the problem. Fortunately, we could do so without inventing a new type, just returning more resource state per item.

See also: *Information Resource* (1.2)

1.4   LINKED DATA

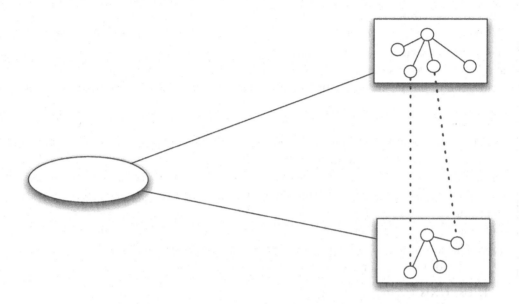

Figure 1.3: Linked Data Pattern.

1.4.1   INTENT

The *Linked Data* Pattern is an approach to exposing data sets with network-friendly technologies. The entities and their relationships are all given identifiers that can be resolved and explored. It also bridges the worlds of network addressability and non-network addressability. Resources that do not directly reside in a machine-processable form are still useful as means of identifying people, places, concepts, or any term or relationship from a domain of interest. Once a data set is made available like this, it becomes possible to connect it to a variety of other data sets, including ones you do not control.

1.4.2   MOTIVATION

The progression we have seen between the two previous patterns was from exposing individual resources, through logical identifiers to collections of these resources. By taking advantage of embedded links and response headers we can imagine building an ecosystem of interrelated content for our organization. The account resources can link to the order resources, which in turn can connect to the product resources. This gives us the ability to follow our noses, use content negotiation to request the information in specific formats, get architectural separation from implementation details, increase the scale of requests horizontally, etc.

As we will discuss in the *Gateway* Pattern (2.3), this shifts much of the burden to the client. Even if there is consistency in the design of the various representations, it requires custom development to integrate one XML schema with another one, or to merge two separate JSON documents. Within an organization or the confines of a particular domain, there is sufficient value to go to the trouble of connecting these resources through custom object models or similar strategy. This absolutely limits the scale of integration though. There is simply not going to be enough economic motivation to do this too many times outside of well-understood and clearly useful collections of datasets.

And, yet, the Web is filled with information on a variety of topics that we would love to be able to connect. This is where linked data comes in. By embracing the infrastructure of the Web and a data model that is designed to accept input from other sources, we can minimize the economic disincentive and demonstrate the value of building Webs of interrelated data, not just documents.

Linked Data as a specific approach grew out of the Semantic Web activity and is usually framed upon its enabling standards such as *URIs*, the *Resource Description Framework*, *SPARQL*, and more.[18] There are several public efforts afoot (we will discuss the main one below), but as an approach, it could just as easily provide value within an organization as well.

The idea can be summarized by the following four rules from Tim Berners-Lee:[19]

1. Use URIs as names for things.

2. Use HTTP URIs so that people can look up those names.

3. When someone looks up a URI, provide useful information, using the standards (RDF, SPARQL).

4. Include links to other URIs so that they can discover more things.

The first rule is about identification and disambiguation using a standard naming scheme. The URI scheme defines an extensible format that can introduce new naming schemes in the future that will be parseable by the URI parsers of today. By standardizing the mechanism through which we identify things, we will be able to tell the difference between everything we care about, including our entities and our relationships. Primary keys and column names are useless outside of the context of the database. We would be unable to accept arbitrary information if we were concerned about collisions in our identifiers. Because we use globally unique schemes, we have the freedom be more accepting of third-party data sources.

The second rule helps reinforce the use of global HTTP identifiers by grounding them in domain name-controlled contexts. You cannot publish identifiers in my domains and vice versa. Additionally, HTTP identifiers are resolvable. The name functions as both an identifier and a handle. If I name an entity and use a resolvable identifier, then you must simply issue a GET request to learn more about it.

---

[18]The idea of connected data obviously transcends these ideas. The Open Data Protocol (OData) initiative is another example. See http://odata.org.

[19]http://www.w3.org/DesignIssues/LinkedData.html

The third rule helps us move beyond more straightforward resources, such as documents and REST services, to include non-network-addressable concepts as well. Individuals, organizations, and terms for concepts in general are not to be found on the Web by resolving a protocol. We would like to be able to still learn about them, however, so following a link and discovering more information is a useful strategy.

We run into a conundrum here, though. How can we tell the difference between a URI reference to a concept and one to a document about that concept? Documents are expected to return 200 responses when you resolve them. How is a client to know that a resource is not network-addressable?

Without getting into the larger arguments, we have two main proposals. One is to use fragment identifiers when naming non-network-addressable URIs. By definition, these are not resolvable on their own. So, to refer to myself, I could use the URL `http://example.com/person/brian#me` while using `http://example.com/person/brian` to refer to a document about me.

A second approach is to use `303 See Also` redirects to alert clients that a non-information resource has been requested. I use this approach with the identifier `http://purl.org/net/bsletten`.[20] If you attempt to resolve that resource, you will be redirected to an RDF document that describes me. The redirection is the hint to the client to the general nature of the resource.

There are advocates and detractors of both approaches. The first provides hints in the URI itself and is useful for collections of terms that are not too large and can conveniently be bundled into a single file. The fragment identifier scheme is used more conventionally, however, to bookmark into named portions of documents such as `http://example.com/story#chap03`. A modern browser that resolved that document would probably move the viewport to show the third chapter if it were marked like this:

```
1    ...
2    <div id="chap03">
3       ...
4    </div>
```

To opponents of the fragment identifier scheme, this is too useful of a tool to give up. Additionally, if the collection of terms and concepts is large, resolving the source document for a single non-information resource can be quite expensive. You get the entire thing even if you only wanted part of it.

The non-fragment identifier scheme works better for a large number of identifiers because each can redirect to its own document. You only download what you want. However, opponents of the 303 approach lament the constant roundtrip resolution process needed to resolve information resources. While there is nothing preventing a server from caching these kinds of redirects with `Cache-Control` and `Expires` headers browser support for the feature remains spotty at best.

---

[20]We will discuss Persistent URLs (PURLs) permanent identifiers in the *Curated URI* Pattern section (3.3).

Whichever naming approach you take, the third rule indicates that we want to learn more when we resolve our entity and relationship identifiers. Tell us about the thing we are asking about. What is it? What is it used for? Where can we find more information.

Finally, the fourth rule allows us to connect one resource to related resources using the same techniques. This keeps the linkage going from resource to resource and allows us to connect arbitrarily complicated webs of data. The fundamental data structure is a graph. We may never have the entire graph in a single location, but we can still resolve the pieces that we need, when we want to. Because the RDF graph model provides the logical structure that supports the *Open World Assumption* (OWA) and a series of standard serializations, we now have most of the basic pieces in place to share and consume information from any source that uses this form.[21] SPARQL provides both a protocol for accessing the data on the server side, as well as standard query language for asking questions of arbitrarily complex graphs. We will discuss SPARQL more in the *Named Query* pattern (1.5) below.

With these tools and ideas, we can now imagine being able to describe and connect all of the different resources in our organization to each other and to external data sources. An account resource could link to an order resource that could connect to a product resource. If we came across a third-party source of reviews for this product and could request or produce RDF representations of those reviews, we can automatically connect them to our existing product data without writing custom software.

RDF is expressed as a series of facts or "triples". A subject is connected to a value through a relationship called a predicate. The subject might be like a key in a database. The predicate is like a column name. The value is obviously like the value in a column for a row in a relational database. As we indicated above, the interesting thing about these facts is that we use URIs to identify the entities and the relationships. When we are talking about Linked Data, we will use HTTP-based identifiers.

If I wanted to express that I know Manu Sporny, I would need an identifier for myself, for Manu, and for the notion of "knowing someone." I have already indicated that I use `http://purl.org/net/bsletten` for myself. There is a collection of RDF terms, classes, and relationships called *Friend of a Friend* (FOAF)[22] that has a term called knows grounded in a resolvable HTTP identifier (`http://xmlns.com/foaf/0.1/knows`). If you try to resolve that term, you will be redirected to a document that describes the meaning behind these terms. Each identifier is expressed in a machine-processable way using RDF.

So, if I know that Manu's URI is `http://manu.sporny.org/about#manu`,[23] I could express that in an RDF serialization called N-Triples:

```
1  <http://purl.org/net/bsletten> <http://xmlns.com/foaf/0.1/knows>
        <http://manu.sporny.org/about#manu> .
```

---

[21]We do not require that a data set be stored in RDF, just that, when resolved or queried, it can provide the information in RDF so that it is easily connected to other content.

[22]`http://foaf-project.org`

[23]Note he uses a fragment identifier where I use a 303 redirect!

The graph representing that fact would conceptually look like this:

Figure 1.4: An RDF graph of one fact.

If I added the following facts about our names:

```
1  <http://purl.org/net/bsletten> <http://xmlns.com/foaf/0.1/name> "Brian Sletten" .
2  <http://manu.sporny.org/about#manu> <http://xmls.com/foaf/0.1/name> "Manu Sporny"
```

I now have three facts about the two of us. Notice that nothing has yet indicated that Manu also knows me. It is a directional relationship. We can engage more technology later to allow these kinds of statements to be assumed to go both directions so we do not have to say both facts. That is outside the scope of this discussion, however.

The new graph would look like this:

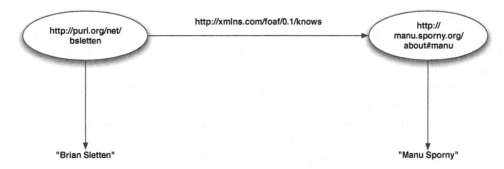

Figure 1.5: An RDF graph of three facts.

One of the standard formats for publishing RDF is called RDF/XML. These three facts serialized in that format would look like:

```
1  <?xml version="1.0" encoding="utf-8"?>
2  <rdf:RDF xmlns:rdf="http://www.w3.org/1999/02/22-rdf-syntax-ns#">
3    <rdf:Description rdf:about="http://purl.org/net/bsletten">
4      <ns0:knows xmlns:ns0="http://xmlns.com/foaf/0.1/"
            rdf:resource="http://manu.sporny.org/about#manu"/>
5    </rdf:Description>
6    <rdf:Description rdf:about="http://purl.org/net/bsletten">
7      <ns0:name xmlns:ns0="http://xmlns.com/foaf/0.1/">Brian Sletten</ns0:name>
8    </rdf:Description>
9    <rdf:Description rdf:about="http://manu.sporny.org/about#manu">
```

```
10      <ns0:name xmlns:ns0="http://xmls.com/foaf/0.1/">Manu Sporny</ns0:name>
11      </rdf:Description>
12    </rdf:RDF>
```

This is the same information as expressed in the N-Triples format, it has just been converted into XML. If I published that file on the Web, anyone could resolve it and discover these three facts. If they were unfamiliar with the identifiers for myself, Manu, or the FOAF knows and name relationships, they would be able to resolve those and learn what was intended by them as well.

If Manu published a similar file somewhere else and used the same names,[24] someone could read that file as well. If both files were read into the same in-memory model or RDF-aware datastore (e.g., Stardog,[25] Virtuoso,[26] AllegroGraph,[27]) then the two graphs would connect automatically. This is the power of RDF. It seems more complicated than simple data formats like JSON and XML, but it also enables some tremendous data integration capabilities, largely for free.

When we apply the four rules from above with these data models, we can start to imagine what a Web of Data would look like. We have barely touched upon strategies for using these ideas more widely. For further guidance on building systems around the Linked Data principles, please see Heath and Bizer [2011].

1.4.3   EXAMPLE

There is a very public initiative called *The Linked Data Project*[28] that applies these ideas to connect publicly available data sets. The people involved with that effort have been tremendously successful in applying these ideas at scale on the Web. They began in 2007 with a small number of data sets that were publicly available.

These data sets were managed by different organizations and available individually. The Geo-Names database is a Creative Commons-licensed collection of millions of place names from every country in the world.[29] DBPedia[30] provides structured representation of much of the human-curated content from Wikipedia. The idea behind connecting them was that the entities from one source could be linked to the other sources in ways that unified the collected facts about the resources. For example, facts about the subject of "Los Angeles, CA" from the GeoNames project could be connected to the facts provided DBPedia, the U.S. Census, etc. While it is easy to imagine connecting the same entities across the data sources, it is also possible to cross domains so that population information can be connected to a geographic region, its crime patterns, job opportunities, education rates, etc. If the integration costs were kept low, the value proposition would not have to be too strong to be worth it. Someone at some point would be interested in navigating that Web of Data.

---

[24]It is not required that he use the same names, but it is certainly more convenient if he does.
[25]http://stardog.com
[26]http://virtuoso.openlinksw.com
[27]http://www.franz.com/agraph/allegrograph/
[28]http://linkeddata.org
[29]http://www.geonames.org/
[30]http://dbpedia.org

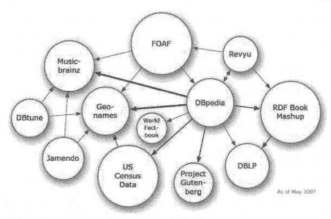

Figure 1.6: The original Linked Data Project data sets.

After an explosive few years, the project has grown to connect an unimaginable amount of data. When you consider how long it takes most organizations to connect a single new data source using conventional technologies, the success is even more pronounced.

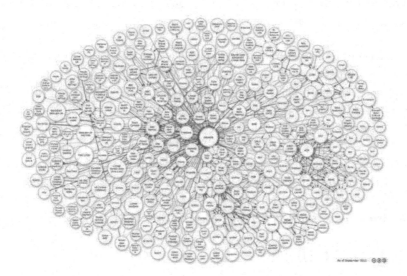

Figure 1.7: Linked Data Project data sets after several years.

Not all of the data is "clean." Some people may not care for the connections or may not trust a source of data. We are collectively building experience at managing these large data sets more carefully to improve the quality, indicate licensing grants, provenance chains, etc.

Tools such as PoolParty[31] are starting to mix natural language processing and Linked Data to generate semi-automated taxonomies of domain-specific terms to help organize content. Services like OpenCalais[32] are using entity extraction and Linked Data to identify references to people, places, organizations, etc. within documents to help structure them. This is only the beginning of how these connected data sources will provide value to us. The *Linked Data* Pattern helps guide us in the use of these techniques for our organizational content as well. We will no longer always be required to write custom software to do integration between data sources that are valuable to us.[33]

1.4.4   CONSEQUENCES

The extra complexity of modeling both the entities and the relationships as URIs has seemed too heavy for many people for years. When compared to the simplicity of a JSON document, or the speed of a SQL query, it is not entirely clear to developers that it is worth it. While it is legitimate to question the burden of a particular pattern or abstraction, the direct approach obscures all of the benefits that RDF and Linked Data provide.

JSON documents cannot generally be merged without writing custom software. Any RDF system can connect just about any other RDF whenever and however it wants to. This is not magic, we are just using global identifiers and a tolerant data model designed to share information. We will see examples of this elsewhere in the book. Additionally, SPARQL queries can run against information kept in arbitrary storage systems (e.g., documents, databases, etc.). As the tools have gotten better, the unnecessary complexity and slowness to process and query claims have become unconvincing.

Until recently, an RDF and SPARQL-based system was going to be mostly read-only, as there was no mechanism to update the results. Additionally, fans of hypermedia and REST bemoan the lack of hypermedia controls in RDF datasets. These issues are largely addressed via SPARQL 1.1,[34] the *Linked Data Platform* Working Group,[35] and the *SPARQL 1.1 Graph Store HTTP Protocol.*[36]

Despite all of these advances, it is still entirely reasonable to want the merged data sets in plain JSON for interaction with popular Web UI frameworks. There is nothing saying that you cannot re-project the data in one of these formats. Not everyone has to touch RDF and LinkedData, even if everyone can benefit from them. Finding the appropriate place to apply any technology is crucial to understanding its consequences.

See also: *Information Resource* (1.2), *Named Query* (1.5) and *Curated URIs* (3.3).

1.5   NAMED QUERY

---

[31]http://poolparty.biz

[32]http://www.opencalais.com

[33]Amusingly, Watson, IBM's Jeopardy-winning system, used information from the Linked Data project to beat its human competitors.(http://www-03.ibm.com/innovation/us/watson/)

[34]http://www.w3.org/TR/2013/REC-sparql11-query-20130321/

[35]http://www.w3.org/2012/ldp/wiki/Main_Page

[36]http://www.w3.org/TR/sparql11-http-rdf-update/

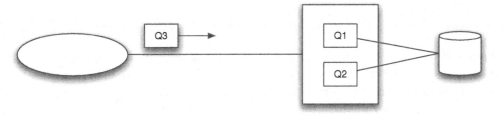

Figure 1.8: Named Query Pattern.

1.5.1 · INTENT

The *Named Query* Pattern generalizes the idea of providing an identifier for a reusable query. The naming process might be explicit or implicit, based on ad hoc, but identifiable, query structures. By giving the query identity, it becomes reusable, shareable, cacheable, and perhaps a new data source.

1.5.2 MOTIVATION

In Sections 1.2 and 1.3, we learned about information resources and collections of information resources. These are generally sets of information uniquely identified by logical names managed by the producers of the information. It is easy to imagine the pre-defined set of resources being insufficient for a client's needs over time. There may be new elements of interest, new ways to organize the content, or even just experimental activity that a user would like to be able to define, bookmark, and perhaps share with others. We would like to do this without requiring software engineers in the middle to capture the requirements, build an endpoint, and deploy the new information sources. That is often simply too slow of a process and is frequently not necessary.

Historically, we have managed this capability by allowing users to run arbitrary queries against our data stores, or extract information into spreadsheets for further manipulation and exploration. The down side of this approach is that it does not usually facilitate the sharing and reuse of this work. We e-mail documents with the results in them to others or push them to a Wiki or other document-management system. Both approaches generally lock up the information in ways that make them less useful.

If it were possible for clients to define queries into the data sources they were interested in and share the results directly, we would get the benefits of user-driven exploration and server-mediated information sharing. This is what the *Named Query* Pattern does for us.

The first approach to this pattern would be for the client to POST a query definition in some agreed-upon format to a known endpoint. If the named query endpoint is http://example.com/query, a client might POST a query to it. The expression of the query will be language and usage specific, but we can imagine it being submitted either in the body of the request or as a URL-encoded form submission. Once the server accepts the request, it approves and creates the endpoint bound to it, and returns the following:

```
1   HTTP/1.1  201  Created
2   Location:  http://example.com/query/id/10
```

From this, the client learns the identity of the *Named Query* resource in the `Location` header in the response.[37] This link can be bookmarked or shared. Anyone interested in the result can simply issue a GET request to fetch the results.

The other main approach for supporting this pattern is to have an endpoint that accepts a `?query=<URL-encoded-query-definition>` parameter. This approach skips the server-created resource step and simply allows the client to specify the query as a URL. There is no resource discovery process, the identity of the requested URL is the resource to request.

1.5.3   EXAMPLE

The SPARQL Protocol[38] is a great example of a standard supporting the *Named Query* Pattern. Any server that wants to support the protocol will provide a resource endpoint such as `http://example.com/sparql`. Clients have the ability to submit queries as:

- GET request and query parameter

- POST request and URL-encoded parameters

- POST request with query in the body

The SPARQL query language[39] has many forms including SELECT, ASK, DESCRIBE and CONSTRUCT. The SELECT form is the most common form and it returns a result set based on matching a graph pattern and selecting bindings between variables and the selected row values.

The DBPedia[40] site makes structured information extracted from Wikipedia available as RDF data. It supports a SPARQL Protocol endpoint (`http://dbpedia.org/sparql`) powered by the Virtuoso engine from OpenLink Software.[41]

Consider the sample query that looks for German musicians who were born in Berlin:

```
1   PREFIX :  <http://dbpedia.org/resource/>
2   PREFIX rdfs:  <http://www.w3.org/2000/01/rdf-schema#>
3   PREFIX dbo:  <http://dbpedia.org/ontology/>
4
5   SELECT ?name ?birth ?description ?person WHERE {
6       ?person dbo:birthPlace :Berlin .
7       ?person <http://purl.org/dc/terms/subject>
            <http://dbpedia.org/resource/Category:German_musicians> .
8       ?person dbo:birthDate ?birth .
9       ?person foaf:name ?name .
```

---

[37]The client should not generally need to be able to define the name of the endpoint, but that is certainly a variation you could support should the requirement arise.
[38]http://www.w3.org/TR/sparql11-protocol/
[39]http://www.w3.org/TR/rdf-sparql-query/
[40]http://dbpedia.org
[41]http://virtuoso.openlinksw.com

```
10        ?person rdfs:comment ?description .
11        FILTER (LANG(?description) = 'en') .
12    }
13  ORDER BY ?name
```

This can be URL encoded into the following query:

> http://dbpedia.org/sparql?query=PREFIX%20%3A%20%3Chttp%3A%2F
> %2Fdbpedia.org%2Fresource%2F%3E%0APREFIX%20rdfs%3A%20%3Chttp%3A%2F%2Fwww.
> w3.org%2F2000%2F01%2Frdf-schema%23%3E%0APREFIX%20dbo%3A%20%3Chttp%3A
> %2F%2Fdbpedia.org%2Fontology%2F%3E%0A%0ASELECT%20%3Fname%20%3Fbirth%20
> %3Fdescription%20%3Fperson%20WHERE%20%7B%0A%20%20%20%20%3Fperson%20dbo
> %3AbirthPlace%20%3ABerlin%20.%0A%20%20%20%20%3Fperson%20%3Chttp%3A
> %2F%2Fpurl.org%2Fdc%2Fterms%2Fsubject%3E%20%3Chttp%3A%2F%2Fdbpedia.org
> %2Fresource%2FCategory%3AGerman_musicians%3E%20.%0A%20%20%20%20%3Fperson
> %20dbo%3AbirthDate%20%3Fbirth%20.%0A%20%20%20%20%3Fperson%20foaf%3Aname
> %20%3Fname%20.%0A%20%20%20%20%3Fperson%20rdfs%3Acomment%20%3Fdescription
> %20.%0A%20%20%20%20FILTER%20(LANG(%3Fdescription)%20%3D%20'en')%20.%0A
> %7D%0AORDER%20BY%20%3Fname.[42]

Which yields the following results:

```
1   <?xml version="1.0"?>
2   <sparql xmlns="http://www.w3.org/2005/sparql-results#"
3       xmlns:xsi="http://www.w3.org/2001/XMLSchema-instance"
4       xsi:schemaLocation="http://www.w3.org/2001/sw/DataAccess/rf1/result2.xsd">
5     <head>
6       <variable name="name"/>
7       <variable name="birth"/>
8       <variable name="description"/>
9       <variable name="person"/>
10    </head>
11    <results distinct="false" ordered="true">
12      <result>
13        <binding name="name">
14          <literal xml:lang="en">Alexander Marcus</literal>
15        </binding>
16        <binding name="birth">
17          <literal datatype="http://www.w3.org/2001/XMLSchema#date">
18            1972-07-26
19          </literal>
20        </binding>
21        <binding name="description">
22          <literal xml:lang="en">Alexander Marcus (born 26 July 1972 in Berlin) is
                a persona of ...</literal>
23        </binding>
24        <binding name="person">
25          <uri>http://dbpedia.org/resource/Alexander_Marcus</uri>
26        </binding>
27      </result>
```

[42]For your convenience, I have created a TinyURL (http://tinyurl.com) for the this long query: http://tinyurl.com/
8hrq4y6.

```
28    < result >
29        < binding name="name" >
30          < literal xml:lang="en" >Andy Malecek</ literal >
31        </ binding >
32        < binding name="birth" >
33          < literal datatype="http://www.w3.org/2001/XMLSchema#date" >
34              1964−06−28
35          </ literal >
36        </ binding >
37        < binding name="description" >
38          < literal xml:lang="en" >Andy Malecek is a German guitarist who has played
                  in the ...</ literal >
39        </ binding >
40        < binding name="person" >
41          < uri >http://dbpedia.org/resource/Andy_Malecek</ uri >
42        </ binding >
43      </ result >
44      ...
45    </ results >
46  </ sparql >
```

One of the more compelling examples of this pattern is mixing the SPARQL protocol with the CONSTRUCT form of the query. Rather than returning the result set shown above, this form of query returns an actual graph.

If we have the following query:

```
1  PREFIX : <http://dbpedia.org/resource/>
2  PREFIX rdfs: <http://www.w3.org/2000/01/rdf−schema#>
3  PREFIX dbo: <http://dbpedia.org/ontology/>
4
5  CONSTRUCT { ?musician ?p ?o }
6  WHERE { ?musician <http://purl.org/dc/terms/subject>
          <http://dbpedia.org/resource/Category:German_musicians>;
7  ?p ?o
8  }
```

We are defining the graph pattern in the WHERE clause like we do for the SELECT form. However, we are constructing a new graph (rather a sub-graph) of all facts we know about anyone who is a German Musician. When mixed with the SPARQL Protocol:

http://dbpedia.org/sparql?query=PREFIX%20%3A%20%3Chttp%3A%2F
%2Fdbpedia.org%2Fresource%2F%3E%0APREFIX%20rdfs%3A%20%3Chttp%3A%2F%2Fwww.
w3.org%2F2000%2F01%2Frdf-schema%23%3E%0APREFIX%20dbo%3A%20%3Chttp%3A%2F
%2Fdbpedia.org%2Fontology%2F%3E%0A%0ACONSTRUCT%20%7B%20%3Fmusician%20
%3Fp%20%3Fo%20%7D%20%0AWHERE%20%7B%20%3Fmusician%20%3Chttp%3A%2F%2Fpurl.
org%2Fdc%2Fterms%2Fsubject%3E%20%3Chttp%3A%2F%2Fdbpedia.org%2Fresource
%2FCategory%3AGerman_musicians%3E%3B%0A%3Fp%20%3Fo%0A%7D[43]

---

[43] Also shortened for your convenience: http://tinyurl.com/ckho3gr.

we have defined a subgraph extraction of DBPedia's much larger data set. We could grab it using curl. By default, the endpoint returns the graph as Turtle.[44] If we want it as RDF/XML, we can use content negotiation to get it:

```
1   curl −L −H "Accept: application/rdf+xml" http://tinyurl.com/ckho3gr >
        musicians.rdf
```

This becomes a data source we can run another SPARQL query against. If we have this query in a file called `sparql.rq`:

```
1   PREFIX dbpedia−owl: <http://dbpedia.org/ontology/>
2   PREFIX foaf: <http://xmlns.com/foaf/0.1/>
3
4   SELECT ?name ?thumbnail
5   WHERE { ?s  foaf:name ?name ;
6      dbpedia−owl:thumbnail ?thumbnail
7   }
```

we will get the name and a link to the thumbnail of any German Musician identified in our subgraph. If we use the Jena[45] `sparql` command line tool to run against our data:

```
1   sparql −−data musicians.rdf −−query sparql.rq
```

we will get a long list of names and URLs to thumbnail images.

If we wanted to extract the subgraph as N-Triples[46] in one go, we could use the `rdfcat` command line tool:

```
1   rdfcat −out ntriples http://tinyurl.com/ckho3gr
```

to get a very long list of RDF triples, one per line.

Ultimately the *Named Query* we defined has become a reusable resource unto itself, which was the entire point of this pattern.

1.5.4   CONSEQUENCES

One positive consequence of this pattern is the ability to cache results of potentially arbitrary queries. The approach where the user submits a query and learns of the created resource is more cacheable because the server can give it a stable name. As long as the client bookmarks it or shares it, subsequent requests can be cached. If the client does not bookmark the new resource, it may be more difficult to find it after the fact. How would you define or describe it? The server could track queries submitted by each user and allow them to return to historical requests.

Associating the query with just the user, however, might complicate the ability to reuse the query across users who ask the same questions. If the query is put into a canonical form, the server might be able to reuse existing links when someone submits equivalent queries in the future.

---

[44]http://www.w3.org/TR/turtle/
[45]http://jena.apache.org
[46]http://www.w3.org/2001/sw/RDFCore/ntriples/

The second approach of passing the query in as a parameter does not require a step to find a related URL to a resource. The downside is that, given the user definition of the query (and thus the URL), it is going to be less likely that the exact same query is asked by multiple users. Renaming or reordering query variables would result in a different URL, defeating the potential for caching between users. Subsequent requests to the exact same *Named Query* resource, due to bookmarks or shared links, can still be cached.

The major negative consequence of this pattern is the potential for clients to impose non-trivial computational burdens on your backend resources. Unconstrained and poorly written queries could potentially consume too much processing. If this is a problem, or you wish to avoid it being a problem, there is nothing that precludes the use of identity-based authentication and authorization schemes to control access and minimize the impact a client can have. Pay special attention to the mechanism by which clients are issuing requests to these resources. A complicated authentication scheme might be overkill. HTTP Basic Authentication over SSL may be sufficient to restrict access, but in a widely supported and low-effort approach.

Beyond simply restricting the *Named Query* Pattern to trusted and registered clients, you may wish to add support for velocity checking and resource use quotas per user. This would reduce the negative impact an individual client can have on the server by throttling access to the backend, based on cumulative use or request rate. The combination of the *Resource Heatmap* and *Guard* patterns can help apply these protections in declarative and reusable ways.

Another concern in allowing arbitrary queries to be submitted by clients is the risk of privilege escalation. Data sources of sensitive information usually have quite rigid access control policies in place, based upon the identity of the user. In supporting open-ended queries, you may want to avoid creating new accounts in your data stores for potentially one-off users. You might consider allowing system or channel-specific accounts to issue queries on behalf of the distributed users. If you have sensitive tables, columns, attributes, and values, you may wish to engage a declarative approach to filter or reject queries that attempt to access them from non-privileged accounts.

See also: *Resource Heatmap* (2.4) and *Guard* (2.2).

CHAPTER 2

Applicative Patterns

2.1   INTRODUCTION

The *Applicative* Patterns introduced in this chapter are not generally primary resources such as those described in Chapter 1. These patterns apply, shape, analyze, and manipulate other primary resources, allowing us to extend them into new uses. We will extract, transform, and derive metadata about things potentially deployed by other providers.

Resource abstractions are powerful ways of sharing and integrating arbitrary information from a wide array of content providers, but they also push much of the burden and complexity to the client. The individual resource producers are not required to coordinate the selection of particular representation formats or interaction styles. In the *Gateway* Pattern, we will see a strategy to broker the relationship between a client and upstream content by introducing a client-side abstraction.

The idea is that a commitment to a large, stable, resource ecosystem will produce innumerable pieces of information for us to find, bookmark, share, and build upon. They will be introduced incrementally, organically and (likely) without any organizing vision. The Information Technology departments of most organizations often feel they need to direct and centralize the flow of information, but the Web has shown us this is not a requirement. Arguably, it may not even be a good idea, given the varied world views and dynamic contexts with which any reasonably sized organization must contend. Therefore, we must be able to apply patterns to manipulate information from existing resources into something we can use.

While there is a *Uniform Interface* in Web architectures, each resource handler will determine which aspects of the interactions to honor. Not all resources will be able to withstand the load of sudden interest from a viral or ascendant source of information. Not all resources will use the same schemes for authentication and authorization, at least not at first. We will want to be able to shape the traffic and protect upstream resources using reusable approaches. We want these secondary resources to benefit from and build upon the primary sources in ways that do not require knowledge of implementation-specific details. In order to achieve all of these different goals, we will use the abstraction to protect the abstraction.

2.2   GUARD

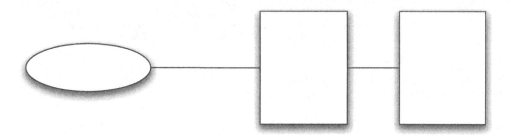

Figure 2.1: Guard Pattern.

2.2.1   INTENT

The *Guard Pattern* serves as a protection mechanism for another resource, service, or data source. Client requests can be funneled through a separate resource to force a specific authentication, authorization, throttling, or other strategy. The ultimate resource handler can usually remain unaware of the guarded concerns. It is a strategy to apply external standardization, consistency, and reuse across potentially independent and perhaps unrelated resources.

2.2.2   MOTIVATION

The concept of interception is widely used in the software development world to wrap a behavior with some modularized functionality you desire to happen first. In most cases, it requires a standard interface of some sort to achieve this goal. The *Decorator* Pattern[1] is an object-oriented strategy that uses a common language-level interface.

As an example, this interface describes some (not very useful) behavior we would like to occur.

```
1   public interface SomethingInteresting {
2       public void beInteresting();
3   }
```

Our inner class implements this interface and does whatever it was designed to do.

```
1   public class Inner implements SomethingInteresting {
2       public void beInteresting() {
3         System.out.println("Doing something interesting.");
4       }
5   }
```

Our decorator class also implements the interface and does its thing before delegating to the inner instance.

[1]http://en.wikipedia.org/wiki/Decorator_pattern

```
1    public class Decorator implements SomethingInteresting {
2        private SomethingInteresting si;
3
4        public Decorator(SomethingInteresting si) {
5          this.si = si;
6        }
7
8        public void beInteresting() {
9            System.out.println("I am interesting before you.");
10           si.beInteresting();
11       }
12   }
```

Finally, elsewhere in the code, we wrap our inner instance and issue a call to the behavior we would like to invoke:

```
1    Inner i = new Inner();
2    Decorator d = new Decorator(i);
3    d.beInteresting();
```

and we get the expected output:

```
I am interesting before you.
Doing something interesting.
```

This approach does allow us to modularize specific behaviors into separate classes, which lowers the complexity and increases the potential for reuse. In this case, the `Inner` class can be used on its own, or in the decorated form. Additionally, the `Decorator` class could be used to wrap other instances of the `SomethingInteresting` interface.

While it is a useful pattern for modularizing the concerns of logging, security checks, managing transactions, managing database connections, etc. around specific behavior, it requires a structural relationship between the decorator and the inner instance. This ends up being a rather heavy code-level dependency. It also limits how the wrapped behavior can be applied in other circumstances. Other interfaces unrelated to this one may also require the same behavior, which would complicate the potential for reuse.

Aspect-oriented programming[2] takes the idea a step further by explicitly supporting a mechanism to intercept and decorate behavior structurally without the use of a specific interface. It allows you to modularize interception behavior of arbitrary functionality by detailing how the decorating concerns can be "woven" in around specific events in the program's lifecycle. You might even want to force the inner class to always be decorated so that it can never be used in isolation.

These ideas have been adopted enthusiastically in software environments such as Spring[3] to allow configuration to select the composition of architectural cross-cutting concerns without having to modify software.

---

[2]http://en.wikipedia.org/wiki/Aspect-oriented_programming
[3]http://static.springsource.org/spring/docs/2.0.x/reference/aop.html

The *Uniform Interface* of the Web allows us to consider a similar approach. We have a common addressing scheme that works across all of our resources. You could imagine applying the mechanics of the *Transformation* Pattern in Section 2.5 to "route" a reference to a resource through another resource:

```
http://example.com/decorator?reference=http%3A%2F%2Fexample2.com
%2Finner
```

The decorating resource can do its thing prior[4] to the invocation of the inner resource. We could also potentially apply the decorating resource to another resource:

```
http://example.com/decorator?reference=http%3A%2F%2Fexample2.com
%2Finner2
```

The problem is that we may not want the inner resource to be accessible without the decorating behavior. If we are talking about actually guarding a resource, then the above pattern will be a Maginot Line[5] defense that could simply be avoided.

Web architecture also makes it easy to support the redirection of resources. We could imagine rewriting the inner request to the decorated request above. Ignoring potential performance problems, the problem is that we may not ultimately be in control of the resource we are decorating and do not want to be messing around at the *Domain Name Service* (DNS) level.

Instead, we will need to bind the *Guard* Pattern behavior more tightly to the resource being protected. The mechanism for managing that will probably be very implementation-specific, but you could imagine a new resource combining a form of the *Informed Router* Pattern (3.4) mixed with network access control to force traffic through a particular channel. The decoration can be applied regardless of the underlying implementation technologies. For more stack-specific solutions, we could use an interception pattern such as Servlet Filters to decorate the handlers in ways that keep them unaware of the wrapped behavior.

The Restlet[6] framework supports direct decoration of other resources through the use of a Guard class:

```
1    // Attach a guard to secure access to the directory
2    Guard guard = new Guard(getContext(), ChallengeScheme.HTTP_BASIC, "Sample
         App");
3    guard.getSecrets().put("scott", "tiger".toCharArray());
4    router.attach("/items", guard);
5    router.attach("/item", guard);
```

2.2.3 EXAMPLE

One of the downsides of most technology stacks used to produce resource-oriented architectures is that the resource abstraction ends once you cross into the environment. Code level handlers are invoked when the Web container maps the external requests to the configured responder. Inside this

---

[4]Or after, decoration can precede, succeed, or replace the inner functionality if it wants to.
[5]http://en.wikipedia.org/wiki/Maginot_Line
[6]http://restlet.org

"code space," software works the same as it does anywhere else. Outside of that space, we see the flexibility, scalability, and evolvability of the Web architecture.

The NetKernel environment[7] we used to demonstrate the patterns in this book brings the resource-oriented abstractions "inside". Everything is addressable via URIs. The environment resolves resource requests based upon the context in which they are requested. Spaces advertise URIs that then get mapped to internal handlers.

Normal Web resources are resolved within the context of a single resolution space (i.e., DNS). If you expose a resource at `http://example.com/account/id/12345`, then DNS will resolve the host that has the URL pattern exposed and mapped to an internal piece of code. In NetKernel, you can have an arbitrary stack of spaces that control the resolution context. Who is asking for what in which circumstance can all control which resources are resolved in the process.

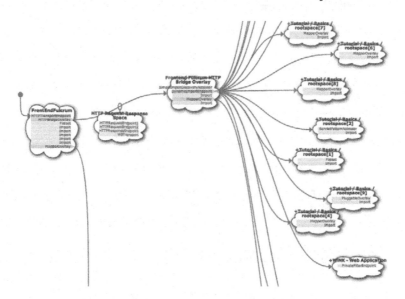

Figure 2.2: Composition of spaces produces architectural features.

What we see in Figure 2.2 is a slice of how NetKernel itself is composed by spaces importing other spaces. The *Front End Fulcrum* (FEF) exposes internal identifiers as HTTP-accessible resources like any other framework would. This includes the NetKernel Wiki application (WiNK) and several tutorials. Internally, however, modules also export URIs. The FEF learns about what to expose via HTTP by importing the spaces of other modules. There is obviously a lot more to what is going on here, but the point is that these are like little DNS look ups (i.e., who is responsible for responding to this URI). Space composition can also be used to restrict access to a particular set of resources (effectively implementing our *Guard* Pattern).

---

[7]`http://netkernel.org`

Everything basically works the same way because the entire system is resource-oriented. Requests are handled by kernel threads asynchronously, which ends up "load balancing" requests across CPUs in the same way a load balancer does to backend resources. Not surprisingly, this ends up demonstrating the same linear scalability we see with Web architectures. Rather than "load balancing" to backend servers, NetKernel does so against CPUs.

What you find is that operational systems end up looking like the basic curve shown in Figure 2.3 regardless of what you are doing:

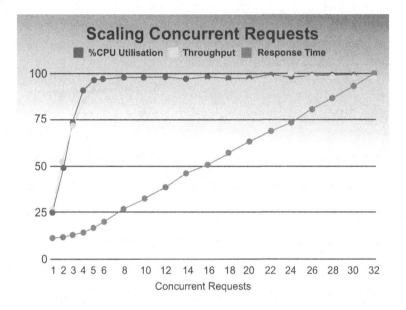

Figure 2.3: Resource-oriented environment displaying properties of the Web.

There are a series of patterns in NetKernel that allow you to compose systems architecturally rather than via code. The *Overlay* Pattern[8] is an example of the ability to insert a resource request in front of another request as a form of wrapped interception. Because everything is resource-oriented, we are able to implement the *Guard* Pattern by binding another resource to the requested resource through an overlay.

Below we see an example from the NetKernel Wiki implementation (WiNK) that overlays the GateKeeper Pattern[9] in front of the space advertising the URIs for the protected resources. These are defined in separate URI-addressable spaces (e.g., urn:org:ten60:wink:www:security:impl), but the effect is to prevent direct access to those resources when coming through this channel. Requests will be rewritten to another resource (configured separately) that will evaluate whatever

---

[8]The *REST Overlay* Pattern is identified here: http://docs.netkernel.org/book/view/book:tpt:http:book:doc:tpt:http:RESTOverlay

[9]http://docs.1060.org/docs/3.3.0/book/cstguide/doc_ext_security_URIGateKeeper.html

is being asked for (conveyed by the `arg:request` argument). Theses could be static files, functional behavior, or references to backend systems. When everything is a resource, we can impose homogenous security policies against anything.

```
1   <space name="Wink — HTTP Session Overlay">
2     <pluggable−overlay>
3       <preProcess>
4         <identifier>active:GateKeeper</identifier>
5         <argument name="request">arg:request</argument>
6       </preProcess>
7       <space name="WiNK — Protected Space">
8         <import>
9           <uri>urn:org:ten60:wink:www:security:impl</uri>
10        </import>
11        <import>
12          <uri>urn:org:ten60:wink:www:core</uri> ·
13        </import>
14      </space>
15    </pluggable−overlay>
16  </space>
```

Another example of the *Guard* Pattern implemented cleanly and simply in this environment is the notion of a *Throttle*. Protecting a backend resource (whether it lives within this environment or externally as an HTTP-based resource) can be as simple as overlaying controlled access to this resource.

Many years ago, I implemented a system in NetKernel for the intelligence community to process millions of documents. It was a complicated pipeline that involved commercial software, open source, and hand-written tools. Documents were accepted on the front end, put into a standard format, run through a series of entity extractors, and then cleaned up. The overall process involved twenty or so steps. Each one was independently addressed and loosely coupled in this environment. Some of the steps had common interfaces, but the majority of them did not, given the disparate technologies involved.

We had early success with the approach, quickly putting together a system that displayed linear scalability when run on servers with extra CPUs. After we had made some progress, however, we were informed by one of the commercial vendors that we had to restrict their software to run on a single thread in order to avoid tens of thousands of dollars of extra licensing costs. We could have solved the problem by implementing the *Decorator* Pattern at the code level, if that particular step had an appropriate interface to use. We could have put a Java `synchronized` block around this interface to stick with the license terms. If, however, we ever needed to purchase additional licenses, even though it was modularized in a decorating class, we would have to move to a *counting semaphore*[10] or other advanced concurrency technique to achieve the different usage pattern. That would have been a disruptive change hidden within the code and only applicable to anything implementing a common interface. It would not have been the end of the world, but fortunately, we had another option.

---

[10]http://en.wikipedia.org/wiki/Semaphore_(programming)

Because each step already had a URI, I was able to use the *Throttle* Pattern to limit access to the obnoxiously licensed component. Based on the following configuration, only one thread would be allowed through at a time. We would be able to enqueue up to 20 requests before rejecting additional requests.

```
1   <overlay>
2     <prototype>Throttle</prototype>
3     <config>
4       <concurrency>1</concurrency>
5       <queue>20</queue>
6     </config>
7     <space>
8       <import>
9         <uri>urn:com:example:extraction:tool</uri>
10      </import>
11    </space>
12  </overlay>
```

This *Guard* Pattern implementation took advantage of the ability to wrap one resource request in another without modifying it. The architectural property of throttling was made explicit in the definition of the pipeline itself. Had we ever needed to purchase an extra license, we could have done so and modified the configuration to allow a second or third thread through without touching any code.

In a non-resource-oriented environment we could still have built a protection mechanism at the code level, but it would have been substantially more complicated. We would have lost the fact that everything is URI-addressable and would have had to rely on code or filter-level patterns. We would also needed to handle the lack of a consistent processing model throughout the environment. We would have had to bridge the fact that external requests were being handled by a set of HTTP-level threads. Blocking one of those inappropriately would lower the throughput of the entire system. It would not have been impossible, but it was made substantially easier by an environment that was resource-oriented through and through.[11]

2.2.4   CONSEQUENCES

The *Decorator* Pattern clearly has consequences in that you need a consistent interface at the code level and some way of wrapping instance objects to prevent calling code from getting directly to the unwrapped object. The *Uniform Interface* of the Web architecture solves some of that for us (everything works the same when wrapped by logical identifiers and accessed by semantically-constrained methods), but we need some way of controlling access to the backend resource we want to guard.

If we are running in an environment like NetKernel where everything is an information resource, then something like the *Guard* Pattern is trivially and non-invasively implemented. If we are running within the same resource context in another environment (e.g., the same instance of

---

[11]For another example of NetKernel throttling being used to optimize the use of cloud resources in a fascinating way, see http://tinyurl.com/bu8pfam.

Restlet or the same Servlet context), we can use code-level composition to protect the resource. If the resource lives on its own in a separate context, we may need to restrict access at the network level to prevent people from resolving the resource directly. Only our resource is allowed to directly talk to it.

By redefining the context in which a resource is accessed, however, we are able to decorate and protect one resource with another.

See also: *Informed Router* (3.4) and *Transformation* (2.5).

2.3   GATEWAY

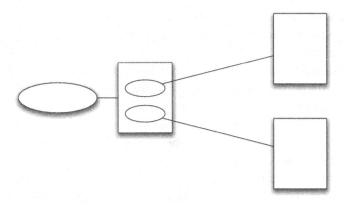

Figure 2.4: Gateway Pattern.

2.3.1   INTENT

The *Gateway* Pattern is notionally a combination of the *Information Resource, Named Query,* and *Transformation* Patterns. The distinction is that it is generally intended for a specific client's use. Rather than a server resource being established for general use, the *Gateway* Pattern establishes a proxy for a client interested in an orchestration, data aggregation, content extraction, or other processing of one or more backend sources. The resource abstraction dependency keeps the specifics of custom software development from impacting the client directly.

2.3.2   MOTIVATION

The REST architectural style provides a series of design decisions that yield desirable architectural properties. It defines a *Uniform Interface* to make the interaction between clients and servers more predictable in networked software environments. It does a great job helping an organization understand how to build scalable information-oriented systems.

What it does not do is make the client's life all that much easier. In most ways, we are shifting the complexity to the client, which ends up becoming more like a browser. It shifts to reacting to what the server tells it, rather than knowing what to expect because the developer coded it to. This breaks the coupling and induces long-term stability and evolvability, but it also is a foreign way of thinking to most developers.

Because resource-oriented ecosystems can grow organically, they usually will. Documents, data, and REST services will pop up incrementally over time. In most organizations, this will proceed apace until someone suddenly realizes that there has been a rapid propagation of "rogue" services. At this point, someone will draft a policy document on standardizing REST services and all productivity will stop until the cycle begins again.[12]

The reality is that these services are never going to be standardized to the extent that most IT organizations would want, nor should they be. The Web grew because someone would publish some content and people would find it and express interest. Within an organization, this is the most likely scenario, too.

The problem remains that clients need to integrate between various data sources and REST services. If they expose information in one XML schema here, another XML schema over there, and then some JSON at a third location, the client will have to be responsible for consuming the different formats and figure out how to connect the data sets. As we discussed in the *Linked Data* pattern, this will be deemed acceptable for small numbers of information sources in relevant domains. It is not a strategy that will scale, however.

Imagine having resources for an account, its order history, the products associated with the orders, information about the manufacturers, the locations of delivery, external product reviews, customer service activity to capture issues, social networking information, etc. If they all used different schemas or representations formats, it would be painful to tie them all together. But, imagine the value of being able to explore and navigate all of it.

If everything were stored in a comprehensive relational database, we could explore the relationships via queries. However, you will never have all of the information in one place like this. Information comes to us from everywhere, including unexpected locations, so we need a strategy that allows us to consume information more readily. One solution, if you are able to influence the source content, is to allow content negotiation of the source content as RDF.

If each of the main types mentioned above were available as RDF (or could be easily transformed into it), then integrating all of the data sources would be pretty trivial. SPARQL queries allow you to ask questions of the resulting structure, even for data sources you may have never seen before. The *Linked Data* Pattern can help you plan strategies to do this.

If that is not an option, then the client will have to come up with its own strategy for integration. One approach might be to create a domain model in a high-level language like Java or C# and build up the state that can be managed over time as new sources are maintained. This is how we have been

---

[12]Why, yes, I have seen this happen more than once, why do you ask?

attempting this for the past two decades, however, and it leaves a lot to be desired. Each new source involves software design, development, testing, and deployment.

The *Gateway* Pattern is an alternate approach to solving the problem. It establishes a client-driven resource abstraction over the various sources. By using dynamic and flexible languages, it minimizes much of the development and deployment burdens. Clients still have dependencies on the resources, but they are protected from being impacted too strongly by schema changes and new content based on language and representation format choices.

2.3.3 EXAMPLE

The ql.io[13] project is a good example of this pattern in the wild. There have been other approaches that have achieved similar results, but this is the most comprehensive and promising example not based on global standards. It takes advantage of JavaScript, Node.js,[14] and the general familiarity people have with relational database tables to define an abstraction over the integration of various resources. Because it represents a comprehensive example of the pattern, we include the following detailed walkthrough from the project to highlight the value of the approach.

The various data sources are treated as tables. Mappings turn the individual XML and JSON representations from the upstream sources into table views. While RDF turns everything into a graph, ql.io turns everything into a table. The expressive power of the one is balanced by the familiarity of the other. In either case, the client has a single composite representation to consume, which shields it from some of the otherwise hairy dependencies.

The Node.js instance runs either locally or deployed on a server somewhere. It can be configured by JSON files stored locally or temporarily through a Web Console running on port 3000. Once the "tables" are defined, users are able to run queries against the instance which yields a JSON result set. Queries that yield especially interesting results can be named as a reusable resource.

If you have Node.js and curl[15] installed, the steps to get a basic example running are as follows:[16]

```
mkdir qlio-app
cd qlio-app
curl https://raw.github.com/ql-io/ql.io/master/modules/template/init.sh |
bash
```

This will fetch the project dependencies and establish a file system structure like this:

```
Makefile
README.md
bin
```

---

[13]Pronounced "cleo": http://ql.io
[14]http://nodejs.org
[15]http://curl.haxx.se
[16]On a Unix-y environment. At the moment, there are issues with Windows support for a few of ql.io's dependencies. When those get resolved, they are planning on supporting it.

```
debug.sh
shutdown.sh
start.sh
stop.sh
config
 dev.json
logs
node_modules
 ejs
 express
 headers
 mustache
 ql.io-app
 ql.io-compiler
package.json
pids
routes
tables
```

Next create a file called `tables/basic.ql` and put the following results into it:[17]

```
1   create table finditems
2     on select get from 'http://svcs.ebay.com/services/search/FindingService/
3     v1?OPERATION-NAME=findItemsByKeywords&SERVICE-VERSION=1.8.0&GLOBAL-ID=
4     {globalid}&SECURITY-APPNAME={apikey}&RESPONSE-DATA-FORMAT=
5     {format}&REST-PAYLOAD&keywords={^keywords}&paginationInput.entriesPerPage=
6     {limit}&paginationInput.pageNumber={pageNumber}&outputSelector%280%29=
7     SellerInfo&sortOrder={sortOrder}'
8     with aliases format = 'RESPONSE-DATA-FORMAT', json = 'JSON', xml = 'XML'
9     using defaults format = 'XML', globalid = 'EBAY-US', sortorder ='BestMatch',
10    apikey = "{config.eBay.apikey}", limit = 10, pageNumber = 1
11    resultset 'findItemsByKeywordsResponse.searchResult.item';
12
13  create table details
14    on select get from "http://open.api.ebay.com/shopping?callname=GetMultipleItems
15      &ItemID={itemId}&responseencoding={format}&appid={^apikey}&version=713
16      &IncludeSelector=ShippingCosts"
17      using defaults format = "JSON", apikey = "{config.eBay.apikey}"
18      resultset 'Item';
19
20  create table google.geocode
21    on select get from
         "http://maps.googleapis.com/maps/api/geocode/{format}?sensor=true
22      &address={^address}"
23      using defaults format = 'json'
24      resultset 'results';
```

This step establishes three tables that are individually populated by issuing GET requests to the given URLs. Note that the parameters can be specified via the with and using clauses. For instance, the API key to invoke the eBay services is contained in the config/dev.json file. By default, there is a key you can use to experiment. For production systems, you should use your own registered key. The mechanism for extracting the results into a table is specified in the resultset clause.

In order to populate the result set, the ql.io backend may need to issue multiple calls to the individual services. Like everything else, this is configurable. Many REST APIs have maximum velocities with which you can invoke them in order to prevent intentional (or unintentional) denial of service attacks.

At this point the tables are defined and you are able to issue a request through the Web Console (http://localhost:3000).

For example, if you issue this command:

```
1  select itemId from finditems where keywords='mini cooper';
```

We are asking ql.io to extract the itemId attributes for the rows in the finditems table, the first table defined above that does an item query based on keywords. Notice that the keywords parameter is passed in from the query, rather than the table definition. If you look in the "Req/resp traces" section at the bottom of the window and click on the "Response" tab, you will see a large result set such as:

```xml
1   <?xml version="1.0" encoding="UTF-8"?>
2   <findItemsByKeywordsResponse
        xmlns="http://www.ebay.com/marketplace/search/v1/services">
3     <ack>Success</ack>
4     <version>1.12.0</version>
5     <timestamp>2012-09-05T00:47:57.031Z</timestamp>
6     <searchResult count="10">
7       <item>
8         <itemId>400318823585</itemId>
9         <title>Mini : Cooper WE FINANCE!! 2007 MINI COOPER S SUPERCHARGED 6SPEED
            SPOILER ONLY 68K</title>
10        <globalId>EBAY-MOTOR</globalId>
11        <primaryCategory>
12          <categoryId>107008</categoryId>
13          <categoryName>Cooper</categoryName>
14        </primaryCategory>
15
16        ...
17
18        <returnsAccepted>false</returnsAccepted>
19        <condition>
20          <conditionId>2500</conditionId>
21          <conditionDisplayName>Certified pre-owned</conditionDisplayName>
22        </condition>
23        <isMultiVariationListing>false</isMultiVariationListing>
24      </item>
25    </searchResult>
26  </findItemsByKeywordsResponse>
```

The resultset is populated from the expressed navigation of
`findItemsByKeywordsResponse.searchResult.item`. Therefore, selecting `itemId` from the
table yields the JSON array:

```
1    [
2      400318823585,
3      140838372329,
4      190719613273,
5      190721054195,
6      170904304842,
7      400318091883,
8      300771979252,
9      230838975262,
10     400318091823,
11     400318821909
12   ]
```

A more complicated query issues requests to geolocate the listings based on the location of
the subselect results and joins them:

```
1    select e.ItemID, e.Title, e.ViewItemURLForNaturalSearch, g.geometry.location
2        from details as e, google.geocode as g where
3        e.itemId in (select itemId from finditems where keywords = 'mini cooper')
4        and g.address = e.Location
```

This yields a new, richer JSON result set:

```
1    [
2    [
3    "400318823585",
4    "Mini : Cooper WE FINANCE!!",
5    "http://www.ebay.com/itm/2007-MINI-COOPER-S-SUPERCHARGED-6SPEED-SPOILER
6        -ONLY-68K-/400318823585",
7    {
8    "lat": 29.7601927,
9    "lng": -95.36938959999999
10   }
11   ],
12   [
13   "140838372329",
14   "Mini : Cooper WE FINANCE!!",
15   "http://www.ebay.com/itm/S-HATCHBACK-TURBO-LEATHER-AUTO-STRIPES-2008-MINI
16       -COOPER-63K-HOUSTON-1-OWNER-/140838372329",
17   {
18   "lat": 29.5589204,
19   "lng": -95.3274022
20   }
21   ],
22   ...
23   ]
```

2.3.4   CONSEQUENCES

As a client-side solution, this pattern is more akin to Unix "Pipes and Filters"[18] solutions than Web resources. As such, the only real consequence is that we may end up not sharing the value created by the effort. If you never share your Unix scripts, we have the same problem. Given that implementations like ql.io do allow you to publish the integrations as new resources, this concern is largely eliminated.

The standards-based approaches described in the *Linked Data* Pattern describe another compelling vision for solving the problem of client-side integration complexity. If the data model there is not an option or appealing to your users, the *Guard* Pattern can be a good alternative.

See also: *Information Resource* (1.2), *Collection Resource* (1.3), *Linked Data* (1.4), *Named Query* (1.5) and *Transformation* (2.5).

2.4   RESOURCE HEATMAP

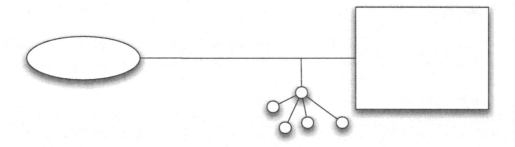

Figure 2.5: Resource Heatmap Pattern.

2.4.1   INTENT

The *Resource Heatmap* Pattern grows out of the dual nature of an addressable resource identifier. As a name, it affords disambiguation in a global information space at a very fine granularity. As a handle, it represents the interface to the resource itself. Metadata captured about how this resource is used, accessed, failure rates, etc. represent crucial business intelligence. This information can be used to identify business opportunities, operational planning, dynamic routing, and more.

2.4.2   MOTIVATION

The Web's initial success at creating a platform for sharing content seamlessly yielded the subsequent problem of how to manage all of the information on it. Early attempts to create directories of the

---

[18]http://www.tutorialspoint.com/unix/unix-pipes-filters.htm

content categorized by humans quickly fell down under the enduring explosive growth and the difficulties of keeping people in the loop. It was simply too painful of a process to be manual.

One of the next great leaps forward was based on the realization that existing linkage between documents could be mined for relevance. A document on a topic that has a large number of inbound links pointing to it is probably more authoritative than one that does not. The value contained in this metadata about the linkage launched the fortunes of Google.

As organizations became more savvy about their own published resources, they engaged elaborate link and traffic analysis platforms to find out which documents people were browsing and how long they were staying on each one. Unfortunately, this information is often produced after the fact and as reports that lock the content up and keep it from being useful in more real-time settings. If the metadata were captured and resurfaced as its own resource, this information could be made available for arbitrary analysis and use throughout the organization. The consistency of the *Uniform Interface* of the Web Architecture and RESTful style gives us a framework upon which to build such a pattern. While it is common to track interest in a resource on the Web, it is less common to do so within the confines of a firewall. A lightweight mechanism that automated most of the process could be quite valuable for assessing the cost of deploying a resource and measuring its use and impact within an organization.

An "overlay," interception, or filtering resource could be applied in the servicing of request handling to keep track of how resources are being used. Request counts, processing times, source IP addresses, consuming applications, client identity, HTTP Referer [sic] linkage, etc. are all examples of the kind of information that could be trivially recorded about a resource. If these were captured and consumed robustly by an organization, it would be easy to begin to identify emerging trends, track the average load, or queue length of shared services, as well as capture where value was being produced. Authorship metadata (who created a resource) could be connected to the usage patterns of the resource as a gauge of who was producing useful information.

Note that just about every web server currently captures some amount of metadata about resource usage in its access logs. The difference here is that we are capturing the information in a structured, reusable fashion and re-projecting it as an information resource in its own right. The *Uniform Interface* of web interactions makes a certain amount of this easy to do, but that is where most access logs stop. What is common to everything is measured, what is specific to certain resources is not. This means we are leaving the majority of the value in an uncaptured state.

One of the most exciting and frightening aspects of the Web is its "anarchic scalability."[19] It is difficult to predict or respond to sudden surges of interest in topics of interest, resource usage, etc. Being able to build an automated mechanism to track this information and feed it back into business systems could shorten the time to awareness of market opportunities, identify unexpected inventory shortages and deployment of new cloud instances to meet bursty needs. The *Informed Router* pattern (3.4) is an example of a way to leverage this information operationally.

---

[19]http://www.ics.uci.edu/~fielding/pubs/dissertation/web_arch_domain.htm

2.4.3 EXAMPLE

`tinyurl.com` was one of the first link shortening sites to gain widespread use. It was built to make it easier to share longer links via e-mail, print, etc. A long link is submitted to the site and turned into a shorter link through the use of a deterministic hash function.[20] As an example, the link `http://www.youtube.com/watch?v=Hfa3d9bpZeE` could be shortened to `http://tinyurl.com/9notafb`. When someone clicks through the shorter link, the service looks up the hashed value `9notafb` as a key and finds the source URL. The client is then redirected to the longer link.

After Twitter exploded to become one of the largest Web communities around, it became an important channel for sharing content. Long URLs would consume precious characters from the 140 character limit, however, which prompted a surge of interest in link shorteners. Sites like `http://bit.ly` started to pop up all over the place because they were convenient, but also because they allowed organizations to track traffic for domains they did not control. The site that expands a shortened URL into its longer form is privy to what is taking off virally, what is losing steam, what has a long steady growth, etc. They are also able to track where clients are coming from to generate this demand. These companies started to engage machine learning and other advanced analysis techniques to discover market and advertising opportunities. Twitter itself eventually stood up a shortener to take advantage of this insight as well as to mediate access to harmful links that were being shared by malicious bots: `http://t.co`.

These services all provide the ability to capture metadata about resource usage and share it for better insight into trends in social media. The same ideas are applicable to an organization's own information production and consumption patterns.

Within the context of information resources you control, it is easy to imagine a reusable capture and publish capability. Most Web frameworks and technology stacks support some notion of a filter or interception capability. By invoking a non-blocking request during the interception process, it would not incur a substantial performance burden to constantly capture this information. The actual implementation of the storage mechanism is not important, but the query and exposure side would benefit from the use of a hypermedia representation. Consider the *Information Resource*, *Collection* or *Linked Data* Patterns.

If someone were to publish a shared resource at `http://example.com/document/id/567890`, the mechanism that responds to resource requests could be modified to capture usage results. Again, this is not a simple access log. It is capturing structured information about a particular resource. The easy things to imagine are who accessed it from what IP address and when. Aggregate request counts are derivable from this, as is some notion of usage.

But what about the less easy stuff? What is the subject of the document? Who wrote it? What customers, markets, or domain-specific references are mentioned in the document? The use of RDFa,[21] Microformats,[22] or natural language processing services such as OpenCalais[23] could

---

[20]`http://en.wikipedia.org/wiki/Hash_function`
[21]`http://rdfa.info`
[22]`http://microformats.org`
[23]`http://www.opencalais.com`

easily be used to discover this type of information as well. See the *Transformation* Pattern (2.5) for more information on how that might work. Once we have this information, we will want to connect it to the results we already have such as request counts, sliding windows of interest, etc.

Why is this important? In part, because of the discoverable, latent value that is presently unavailable to us. If, within your organization, you have an employee who is consistently producing high-value content on a particular subject, this pattern can help surface that information in a usable way. Subject-matter expertise management is a huge problem in large organizations. Putting the right person on the right project is a crucial activity for maximum value and efficiency.

If, there is a widely read or cited document about a customer, or a manufacturing plant, or an industry partner, or other concept relevant to your daily business activities, this is a key mechanism to help value bubble up to a usable place within the organization. Interest in a resource is a valuable property to track. Finding important things is important. Being able to passively measure interest in arbitrary resources in domain-specific ways is absolutely a key tool in the 21st century. Being able to do so over time to detect emerging or waning topics of interest makes it a crucial differentiator in a competitive marketplace. Being able to track it back to the individuals or portions of the organization that are responsible for producing it is a vital means of talent retention.

The key to the process is that the organization discovers what is interesting, it does not predict it. We are not talking about writing document-specific parsing or handling. We are looking to use industry standards and the *Uniform Interface*, as well as data models that are amenable to arbitrary integration. This is the only way something like this is going to work. Allowing employees and users to interact with the content through ratings, reviews, tags, etc. only serves to enrich the metadata capture and linkage process.

Once you measure and integrate this information, you must surface it. This publishing of unexpectedly captured data also requires the use of standards and consistent interactions. We simply do not have the time or resources to prescribe what is important.

To share activity for the above resource, we can imagine a *Heatmap Resource* endpoint such as `http://example.com/stats/document/id/567890` being produced automatically by the system. As new documents are published, their derived resources will flow through. We certainly could use an XML-based hypermedia system to describe the data:

```
 1   <resourceReport href="http://example.com/stats/document/id/567890">
 2       <summary>
 3           <totalRequest>123</totalRequest>
 4           <averageDailyRequest>23</averageDailyRequest>
 5           <lastAccess>2013-03-31 12:52:19</lastAccess>
 6           <author href="http://example.com/employee/id/1832">
 7               <name>Francis Black</name>
 8           </author>
 9       </summary>
10       <keywords>
11           <keyword href="http://example.com/stats/key/andalusia">Andalusia</keyword>
12           <keyword href="http://example.com/stats/key/chien">Chien</keyword>
13           <keyword href="http://example.com/stats/key/film">Film</keyword>
14       </keywords>
```

```
15    </resourceReport>
```

Here we see discoverable, hypermedia links not simply for the report itself, but other resources. If this content mentions a keyword Andalusia, we can discover a link back to a resource that will produce links to other documents with the same keyword. We are able to "follow our nose" from document to keyword to related documents to who authored the content, etc.

This resource could also be content-negotiated into JSON for easy display in a web user interface framework such as JQuery. If we wanted to really embrace RDF and the *Linked Data* Pattern, we can imagine ever more connectivity between the stuff our organization produces and the environment in which we operate. Connecting from the term Andalusia to the DBpedia entry on the region allows us to sort the results based upon demographics, income levels, or other economic indicators available from external data sources.

2.4.4   CONSEQUENCES

The most substantial concern about capturing this information for arbitrary resources is that the cost of capture and retention would outweigh the benefit from doing so. You should consider prioritizing adoption based upon tangible business-value goals before over deploying the pattern. Finding opportunities to automate the capture of manually collected results and getting the information into business customers hands faster is an obvious target. Keep in mind that the goal of the pattern is not to simply capture and store the metadata about resource usage, but to turn it into its own browsable, bookmarkable, linkable resources. These in turn can be tapped to feed further business processing.

See also: *Information Resource* (1.2), *Collection Resource* (1.3), *Linked Data* (1.4), *Informed Router* (3.4)

2.5   TRANSFORMATION

2.5.1   INTENT

The *Transformation* Pattern is a generalization of an approach to producing new resources that extract content from or transform the shape of content from another resource. As with the *Named Query* Pattern, doing so can induce reusability and cacheability from sources that otherwise do not support such properties. It can also be used to add content negotiation to sources that do not otherwise provide it.

2.5.2   MOTIVATION

Our approach to information resources is often to think about them as independent, single sources of content. Each interaction is a stateless request that may spawn other interactions as part of an orchestration, but at any given point it is one client communicating with one information resource. We potentially pass arguments as part of the request and generally react to whatever is returned.

One of the compelling things about using URLs to address resources on the Web, however, is that we can pass a reference to a resource somewhere else. Anyone we pass the reference to is free to

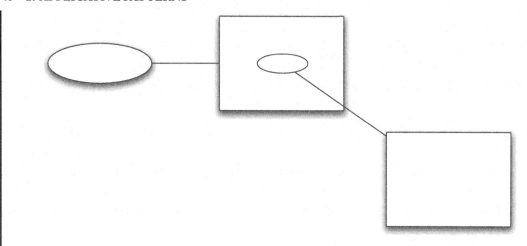

Figure 2.6: Transformation Pattern.

resolve it whenever they want. If an intermediary processor simply passes the reference on to another endpoint, it never even needs to see a representation of the resource. If we are referencing large or sensitive information, this simplifies our architectures to avoid unnecessary exposure or burdens in resolving large messages not needed by intermediate steps in a sequence.

This idea of passing a reference to one resource to another yields the possibility of a compound identifier:

`http://example.com/resource1?reference=<resource2>`.

Obviously, to be a valid URL in and of itself, the referenced resource must be URL-encoded. But the idea of a compound resource creates a compelling way to think about the original resource being transformed somehow. The referenced resource may even point to something out of our control, but that does not prevent us from applying an extraction or conversion in the process.

While it would be easy to go overboard, there is no reason to stop with a single resource reference as long as you are referring to reasonably short URLs.[24] With that in mind, consider the URL:

`http://example.com/resource1?reference1=<resource2>&reference2=`
`<resource3>`.

Clearly this is about as many references as we would want to pass in as part of the URL,[25] but it is enough to start to imagine some compelling uses. Consider a legacy resource that produces an XML representation of some information we cared about. If we wanted to expose the XML as browsable HTML, we could deploy an endpoint somewhere that was configured to point to the

---

[24]There is no standard maximum length for a URL, but common wisdom suggests that anything over 2,000 characters will cause trouble with older Web infrastructure elements. See: `https://www.google.com/search?q=max+url+length`

[25]If you need to submit more resource references, you may wish to put the URL collection in a resolvable resource and submit a reference to that to the *Transformation* Pattern.

XML resource and transform it with an XSLT stylesheet. For a single legacy resource, this would be a reasonable solution. What if we had several such resources? We certainly do not want to have to involve developers for every instance of this kind of transformation.

We could imagine a generic transformation endpoint that could be configured to accept a list of resources and their corresponding stylesheets. Each specific endpoint would look up into the map, find the source resource and its stylesheet, and return the transformed resource. While this is an improvement over having a separate endpoint for each source, it still would involve deployment of new endpoints or maps to extend or modify the configuration. That seems a little heavyweight and inflexible.

If we instead imagine a generic transformation resource that accepts a reference to the XML source and a reference to a stylesheet resource, then anyone could ask for a new transformation at any point. Different users could request the same source resource to be styled via custom stylesheets on demand.

```
http://example.com/transformation?source=http%3A%2F%2Flegacy.example.
com%2Fsome%2Fdoc.xml&xslt=http%3A%2F%2Fexample2.com%2Fstylesheet%2Fsnazzy.
xslt
```

An interesting side effect of this approach is that HTML produced could be cached independently of the other two inputs, but, in practice, you would probably want to reflect changes upstream in the transformed result.

2.5.3   EXAMPLE

One of the more compelling things happening on the Web these days is the steady adoption of ever more structured metadata in the documents being published. We used to try to extract structure from webpages via PERL scripts, but these were generally fragile and the wrong approach. A browser does not need custom parsing for every document it comes across, why should extracting content?

There are several schemes for encoding information in documents including Microformats,[26] HTML5 Microdata,[27] and RDFa.[28] These all represent standard ways of encoding arbitrary domains into documents. Consequently, there are standard ways of extracting the content through standard parsers aware of a particular encoding syntax.

As an example, consider a snippet of information from a page about a conference. In an attempt to draw interest to who will be giving talks, the organizer will highlight the speakers, books they have written, etc. We would expect to see HTML that includes elements such as this:

```
1  <li>
2    <a href="http://uberconf.com/conference/speaker/michael_nygard">Mike
        Nygard</a>, Author of "Release It!", Twitter: <a
        href="http://twitter.com/mtnygard">@mtnygard</a>
3  </li>
```

---

[26]http://microformats.org
[27]http://en.wikipedia.org/wiki/Microdata_(HTML)
[28]http://rdfa.info

Visitors to the page would be able to click through the link to learn more about the author or browse his Twitter account profile. Using modern, standard, structured metadata, however, there is so much more we could do. With RDFa 1.1 Lite,[29] we could imagine the following markup instead:

```
1  <ul prefix="dcterms: http://purl.org/dc/terms/
2            sioc: http://rdfs.org/sioc/ns#
3            foaf: http://xmlns.com/foaf/0.1/">
4    <li resource="http://uberconf.com/conference/speaker/michael_nygard#mike">
5      <span property="foaf:name">Mike Nygard</span>, Author of <span
          resource="urn:ISBN:0978739213" rev="dcterms:creator">"Release It!"</span>
          Twitter: <a property="sioc:UserAccount"
          href="http://twitter.com/mtnygard">@mtnygard</a>
6    </li>
7  </ul>
```

The metadata is woven into the document using a standard approach that makes it easy to extract without having to be familiar with the structure of the document. Any RDFa 1.1 Lite-compliant parser can do so. One of the original ones, PyRDFa, by Ivan Herman, is available either for download or as a service on the web: http://www.w3.org/2012/pyRdfa/.

If you go to that link, you will find a series of options to control the output. These include the means to locate the input document (e.g., via URI, file upload, as pasted text) and the desired output format (e.g., RDF/XML, Turtle, etc.). These options can also be selected based on input parameters to a URL-based extract mechanism that demonstrates our *Transformation* pattern:

http://www.w3.org/2012/pyRdfa/extract?uri=http%3A%2F%2Fexample.com
%2Fuberconf.html&format=turtle&rdfagraph=output&vocab_expansion=false&
rdfa_lite=false&embedded_rdf=true&space_preserve=true&vocab_cache=true&
vocab_cache_report=false&vocab_cache_refresh=false

which yields the following RDF/XML:[30]

```
1  <rdf:RDF>
2    <rdf:Description rdf:about="urn:ISBN:0978739213">
3      <dc:creator>
4        <rdf:Description rdf:about
5          ="http://uberconf.com/conference/speaker/michael_nygard#mike">
6          <sioc:UserAccount rdf:resource="http://twitter.com/mtnygard"/>
7          <foaf:name>Mike Nygard</foaf:name>
8        </rdf:Description>
9      </dc:creator>
10   </rdf:Description>
11 </rdf:RDF>
```

or this Turtle:

```
1  @prefix dc: <http://purl.org/dc/terms/> .
2  @prefix foaf: <http://xmlns.com/foaf/0.1/> .
3  @prefix sioc: <http://rdfs.org/sioc/ns#> .
4
```

---

[29] http://www.w3.org/TR/rdfa-lite/
[30] Namespaces have been elided to make it easier to read in print.

```
5   <urn:ISBN:0978739213>  dc:creator
        <http://uberconf.com/conference/speaker/michael_nygard#mike> .
6
7   <http://uberconf.com/conference/speaker/michael_nygard#mike> sioc:UserAccount
        <http://twitter.com/mtnygard>;
8       foaf:name "Mike Nygard" .
```

The importance of this for our current purposes is that this compound URL to a service that transforms the content from a referenced source becomes a new source of information. We could, for example, use that URL as the source of a default graph in a SPARQL query such as:

```
1   PREFIX foaf: <http://xmlns.com/foaf/0.1/>
2   PREFIX sioc: <http://rdfs.org/sioc/ns#>
3
4   SELECT ?name ?account
5   FROM <http://www.w3.org/2012/pyRdfa/extract?uri=http%3A%2F%2Fexample.com%2F
6     uberconf.html&format=turtle&rdfagraph=output&vocab_expansion=false&rdfa_lite=false
7     &embedded_rdf=true&space_preserve=true&vocab_cache=true&vocab_cache_report=false
8     &vocab_cache_refresh=false>
9
10  WHERE {
11      ?s foaf:name ?name ;
12         sioc:UserAccount ?account .
13  }
```

While someone would not likely issue this kind of query by hand, the idea that you can passively extract content in a machine-processable way from arbitrary resources is a compelling vision of the future of the Web. In this particular case, we could identify an author's social media accounts automatically in order to start following him. If the times and dates for the conference schedule were also expressed this way, we could ask the user if she would like her calendar updated. One resource can be used to transform another resource into a new source of information upon which we might take action.

We can take this a step further. Recall from the discussion of the *Gateway* Pattern in (2.3) that a downside to the REST architectural style is pushing the complexity of data integration to the client. In order to integrate the content in the representations of two or more resources, the client would have to know how to parse the result. Even the use of a simple type like JSON will require the client to know about the domains under question. Consider a resource that is returning information about a book about English royalty. Does the key "title" refer to the book or the honor afforded to a member of royalty?

A data model like RDF makes this kind of integration trivial. A SPARQL query expressed against two different RDF-aware data sources could be as simple as:

```
1   SELECT ?s ?p ?o
2   FROM <URL 1>
3   FROM <URL 2>
4   WHERE {
5       ?s ?p ?o
6   }
```

The data is automatically merged because we use global identifiers for our entities and relationships in RDF. There is no real concern about collisions.

Of course, not everyone in the organization is ready for RDF. Still, the use of RDFa[31] can give us an incremental approach. Existing clients not able to handle RDF can deal with the XML or JSON directly. They will have to continue to know the details of the representations as they do now, but their tools will still work. However, those with "eyes to see" will be able to extract the information from the representation without having to know any details. It seems like magic, but it simply requires the adoption of sufficiently advanced standards.

Imagine we have a hypermedia XML representation of a person:

```
1   <person>
2       <link rel="school" href="http://example.com/school/id/248"/>
3       <name>Brian Sletten</name>
4       <dob>05-26</dob>
5   </person>
```

This representation connects it to another resource representing a school:

```
1   <school>
2       <link rel="admissions" href="http://example.com/admissions"/>
3       <name>College of William and Mary</name>
4       <established>1693</established>
5   </school>
```

If we wanted to find the name of the person, the name of his school, and when it was established, we would have to write some custom code to integrate the two data models.[32] It is unlikely that it would be worth it for a simple need like this.

If, however, our resources were simultaneously encoded with RDFa:

```
1   <person resource="http://example.com/person/id/323"
2       prefix="foaf: http://xmlns.com/foaf/0.1/
3               ex: http://example.com/ns#"
4       typeof="foaf:Person">
5       <link rel="school" property="foaf:schoolHomepage"
6           href="http://example.com/school/id/248" typeof="ex:School"/>
7       <name property="foaf:name">Brian Sletten</name>
8       <dob property="foaf:birthday">05-26</dob>
9   </person>
```

and

```
1   <school resource="http://example.com/school/id/248"
2       prefix="foaf: http://xmlns.com/foaf/0.1/
3               ex: http://example.com/ns#" typeof="ex:School">
4       <link rel="admissions" href="http://example.com/admissions"/>
5       <name property="ex:schoolName">College of William and Mary</name>
6       <established property="ex:established">1693</established>
7   </school>
```

---

[31]Or a Linked Data aware format such as JSON-LD (http://json-ld.org).

[32]Assuming we did not want to go to the trouble of setting up an instance of the *Gateway* Pattern as described in (2.3).

then we simply need an instance of the *Transformation* Pattern to extract the RDF using a standard RDFa parser.[33] Now, a standard SPARQL query:

```
1   prefix  foaf:  <http://xmlns.com/foaf/0.1/>
2   prefix  ex:  <http://example.com/ns#>
3
4   SELECT ?name ?school ?established
5   FROM <http://example.com/transformation/rdfa;resource=http%3A%2F%2Fexample.com
6         %2Fperson%2Fid%2F323>
7   FROM <http://example.com/transformation/rdfa;resource=http%3A%2F%2Fexample.com
8         %2Fschool%2Fid%2F248>
9   WHERE {?student foaf:schoolHomepage ?s ; foaf:name ?name .
10        ?s ex:established ?established ; ex:schoolName ?school }
```

can yield the results we seek:

name	school	established
Brian Sletten	College of William and Mary	1693

text table

Figure 2.7: SPARQL results from transforming RDFa-enabled endpoints.

The goal is not to create custom *Transformation* Pattern endpoints. Instead, we would like to define a reusable pattern through standard identifiers, a *Uniform Interface*, semantically meaningful interaction methods, standard data models, standard encoding mechanisms, standard decoding mechanisms, and a standard query language in order to connect and explore arbitrary domains. That is a remarkable achievement and is eminently doable with the technologies we have discussed here.

2.5.4 CONSEQUENCES

There are no real negative consequences to this pattern, mostly just a need to meet content expectations and perhaps coordination of identity management for disparate resources. Publicly accessible documents that are able to produce structured representations of the information they contain should largely be able to exist without consideration of who is applying these kinds of transformations to them. The source resources should employ a good strategy for alerting consumers to updates via Etags and Cache-Controls. This will assist the transformational resources to know when their extracted or modified content needs to be updated.

Shifts in how a resource is structured will obviously have upstream consequences. A change from Microformats to RDFa, or vice versa, is likely to break any transforming resources that were deployed to produce extracted content. Awareness of who is consuming resources you have published is important in general, not just in the context of this pattern.

---

[33]Several exist in different languages. Please consult `http://rdfa.info/tools/` for more information.

If the source resource requires special credentials to access, the transformer will have to be able to confer those in the process of requesting the source. This situation probably warrants a careful consideration of the applicability of the pattern, however. As long as the chain of authentication is standardized or compatible, this should be a suitable thing to still do. However, any bridging of identities or attempt to "free" controlled content from a protected barrier is probably a big mistake. Security tools such as OAuth[34] could be employed to broker controlled access to referenced resources. Access can be granted to specific applications written by registered developers while still requiring the resource owner's permission to use in a particular context.

[34]http://oauth.net

CHAPTER 3

Procedural Patterns

3.1   INTRODUCTION

The first two chapters of this book introduced primary and secondary resource patterns. The abstraction allows arbitrary information sources to be addressed and resolved through a *Uniform Interface* in the Web architecture. They can be manipulated and transformed into various useful forms through a rich dialogue between the client and server participants without special domain knowledge. These represent relatively straightforward extrapolations of the original implementation of the Web within the context of evolving distributed information-sharing environments.

Even though we can bridge and mask wide implementation variance behind logically named resources, point-to-point and synchronous communication are simply only part of the realities of modern organizational interactions. With the *Procedural Patterns* in this chapter, we now begin to think about requests that occur within a larger ecosystem of related services, workflows, and business processes. We break out of the idea of only using blocking responses and allow disconnected but interrelated subsystems to interact in new ways.

On the surface, it may be hard to imagine how to integrate information resources with asynchronous processes and complicated workflows. By no means do we imagine that all interactions, services, and orchestrations can be easily shimmed behind a resource, but the idea extends further than many people believe. Having a consistent addressing scheme, loosely coupled participants, uniform interfaces, and a predictable infrastructure are part of what the Big Deal is all about. It behooves us to make an attempt to extend the ideas, but not beyond where they are useful.

The first pattern, *Callback*, acknowledges a short-coming in the uni-directional HTTP contexts we have seen for years. On the Web, clients reach out to servers. Updates have generally been delivered via polling. At scale, this is an expensive process. This pattern helps other participants reach back to alert us of updates to application state or progress in a workflow.

Next, we acknowledge our shared history of shame for producing short-lived resources encumbered by bad, fragile, and easily disrupted names. If we are to encourage dependency by time-sensitive and revenue-generating aspects of the business, we must take seriously the responsibility to maintain stability over time. The idea of curating identities into long-lived, reusable resource identifiers is crucial to meet the stability requirements of real businesses. The *URI Curation* Pattern will assist us in doing this.

One aspect of our Web architecture that does not get as much attention is the idea of intermediary processors. As a stateless, visible exchange of self-describing messages, it is useful to inject participants between the clients and origin servers to offload processing hierarchically and at the edge. Caching proxies and load balancers are examples of these processors. Given that they focus upon the *Uniform Interface*, however, there are limits to the kind of operational decision-making they can make. If we extend the concept to include a context-aware routing mechanism, we can make smarter, more adaptive routing decisions based upon application and system state. The *Informed Router* Pattern provides this capability.

Our final item, the *Workflow* Pattern, allows us to embrace the affordances of Hypermedia to affect a client-navigated, coordinated path through a system. While we may not wish to specify a particular user interface experience in order to support a heterogenous client base, we may need to enforce a resource-based workflow. This pattern assists us in doing so.

The larger message of this section is that this resource-oriented abstraction can participate in dynamic and realistic business environments.

3.2 CALLBACK

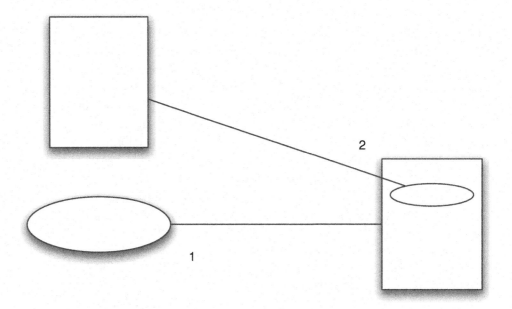

Figure 3.1: Callback Pattern.

3.2.1 INTENT

The *Callback* Pattern is an attempt to route around the limitations of the uni-directional, request/response paradigm of HTTP interactions in resource-oriented systems. By registering or providing links as part of an interaction, the "client" can provide the "server" one or more resources to use to communicate updates, error conditions, or other state changes. In essence, we are simply reversing the relationship between the components. In doing so, we allow resource-based systems to interact with asynchronous, messaging and other types of environments.

3.2.2   MOTIVATION

There is a common misconception that it is impossible to do asynchronous communication using resource-oriented architectures. This is simply not true. HTTP defines a 202 response code that allows a server to accept a request without acknowledging success or completion. As an example, a resource-based order-processing system could accept a representation of an order, return a 202 response code, and then put the order in a queue, send a message, or engage whatever existing backend process is in place.

```
1   POST /request
2   Host: example.com
3   Content-Length: 790
4   Content-Type: application/vnd-order+xml
5
6   <order>
7     <customer>...</customer>
8     <items>
9        <item>...</item>
10       <item>...</item>
11    </items>
12  </order>
```

```
1   HTTP/1.1 202 Accepted
2   Location: http://example.com/status/request/id/12345
```

The client discovers in the response a location to check for updates.

```
1   GET /status/request/id/12345
2   Host: example.com
```

While nothing has changed, conditional GETs, Etags, Cache Controls, etc. can be used to reduce the impact of checking periodically. When things have completed, failed, or updated, the client can be notified the next time they request the status resource.

```
1   HTTP/1.1 200 Ok
2   Content-Length: 206
3   Expires: Wed, 27 Mar 2013 12:15:00 GMT
4   Date: Wed, 27 Mar 2013 12:14:00 GMT
5   Content-Type: application/vnd-async-process+xml
6   Server: NetKernel [1060-NetKernel-SE 5.1.1] - Powered by Jetty
7   Cache-Control: max-age=600
8   Last-Modified: Wed, 27 Mar 2013 12:13:00 GMT
9   Etag: "221207d"
10
11  <process resource="http://example.com/status/request/id/12345">
12      <status>pending</status>
13      <lastUpdate>2013-03-27</lastUpdate>
14      <message>We are currently processing your order.</message>
15  </process>
```

While asynchronous processing is possible, prior to *Web Sockets* and the multiplexed, bi-directional nature of *HTTP 2.0*, the interaction remains a request/response mechanism. The client initiates requests to the server, the server responds. The server does not reach out to the client. This greatly constrains the nature of the interactions possible and introduces a difficult decision. If the client makes the request too often, it puts undue burden on the server and may affect scalability. If it does not issue the request often enough, the system may not be responsive in the face of state changes.

By being willing to accept requests itself, the client can solve this problem by registering a URL for notification when the order is handled. It does not need to poll. The server could even submit periodic updates.

```
1   POST /request
2   Host: example.com
3   Content-Length: 862
4   Content-Type: application/vnd-order+xml
5
6   <order>
7     <link rel="statusResponse" href="http://client.example/status/request"/>
8     <customer>...</customer>
9     <items>
10        <item>...</item>
11        <item>...</item>
12    </items>
13  </order>
```

3.2.3 EXAMPLE

One of the earliest uses of the *Callback* pattern was in *Webhooks*. The term was coined by Jeff Lindsay in 2007. Generally, some workflow or lifecycle event occurs to trigger an HTTP POST to registered URLs from interested clients. The pattern is currently supported by sites such as WordPress,[1] GitHub,[2] MailChimp,[3] and PaperTrail.[4]

As an example, a GitHub user can go to the Settings page of her repository and click on the "Service Hooks" setting. From here, she is able to register Webhooks with over 100 different sites. Any time someone commits something to her repository, these hooks will fire, allowing for integration to a variety of code scanning, project and issue management, and similar applications. The payload of what is sent looks like this JSON template:

```
1   {
2     :before    => before,
3     :after     => after,
4     :ref       => ref,
5     :commits   => [{
6       :id        => commit.id,
```

---

[1]http://en.support.wordpress.com/webhooks/
[2]https://help.github.com/articles/post-receive-hooks
[3]http://apidocs.mailchimp.com/webhooks/
[4]http://help.papertrailapp.com/kb/how-it-works/web-hooks

```
7      :message      => commit.message,
8      :timestamp    => commit.committed_date.xmlschema,
9      :url          => commit_url,
10     :added        => array_of_added_paths,
11     :removed      => array_of_removed_paths,
12     :modified     => array_of_modified_paths,
13     :author       => {
14       :name   => commit.author.name,
15       :email  => commit.author.email
16     }
17   }],
18   :repository => {
19     :name         => repository.name,
20     :url          => repo_url,
21     :pledgie      => repository.pledgie.id,
22     :description  => repository.description,
23     :homepage     => repository.homepage,
24     :watchers     => repository.watchers.size,
25     :forks        => repository.forks.size,
26     :private      => repository.private?,
27     :owner => {
28       :name   => repository.owner.login,
29       :email  => repository.owner.email
30     }
31   }
32 }
```

Another example of the *Callback* pattern is in the OAuth 2.0 authorization flow. The Google Tasks API supports a callback handler to be registered for third-party applications. In addition to registering client IDs and client secrets, through the Google APIs Console, you can register a "Redirect URI" which will be invoked during the authorization process.[5] If your application is hosted at `http://example.com` then it needs to respond to requests at `https://example.com/oauth2callback?code=yis1989`. Other callback URLs in the OAuth 2.0 workflow include support for error handling notifications.

3.2.4   CONSEQUENCES

The obvious issue with this pattern is the requirement for the client to be reachable via the network. One of the reasons the Web has scaled so successfully in a variety of environments is that the clients are not being contacted directly. It is relatively straightforward to open up outgoing ports (e.g., 80, 8080, 443) in a constrained-way, while rejecting all inbound requests. As clients can now include mobile computers, tablets, and smartphones that hop from network to network, a long-running, asynchronous request might not even know where the "client" is at the point that the request is fully processed.

Modern protocols that do allow bi-directional communication are triggered off of a client-initiated request which can make a big difference at the network level. With *WebSockets*, a standard

---

[5]https://developers.google.com/google-apps/tasks/oauth-authorization-callback-handler

HTTP network connection is transitioned into a web socket connection so that the server can send messages back to the client.

Fans of messaging systems may feel this pattern steps into their world unnecessarily. While asynchronous message queues were all the rage a few years ago,[6] the downside was always the need for additional systems and infrastructure to manage the queues. Given that an increasing number of organizations are already exposing HTTP resources to their clients for REST APIs, this pattern works with the existing infrastructure, security policies, etc. As the conversations were already being managed via HTTP on the inbound requests, it seems silly to require a different technology to communicate with the clients if they were willing to accept the HTTP traffic.

3.3 CURATED URI

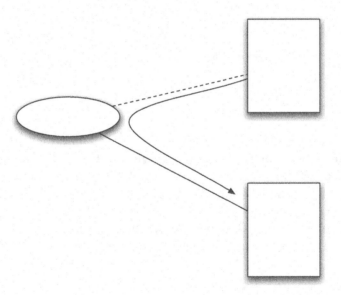

Figure 3.2: Curated URI Pattern.

3.3.1 INTENT

The *Curated URI* Pattern establishes a stewardship commitment to the clients of a named resource. Rather than relying on the good graces of a resource provider to keep an identifier stable over time, a new identifier is chosen to be redirected toward another endpoint. The target resource can be moved, as long as the redirection is maintained, without having a negative impact on clients.

---

[6] And are still widely used today.

3.3.2   MOTIVATION

One of the main goals in a resource-oriented environment is to pick good, long-lived identifiers so that the references to those resources remain stable over time. There will be no resource ecosystem if the names change constantly because it will not be in any client's interest to have dependencies upon unstable sources. As much as we would like this to remain a universal goal among resource providers, chances are that things will break at some point, whether unintentionally or not.

One way to manage this issue is to use an intermediate, purely stable identifier that gets mapped to the actual resource. There can be a curatorial process around the management of these resources. While it should be the goal of a resource provider to be stable, it is ultimately in the client's interest to pursue this property. The good news is that the *Curated URI* Pattern can be managed by either side of the relationship. If a client has interest in a resource but is unsure about its long-term location or stability, she can establish a Curated URI to point to the existing resource and then put her dependency on an identifier she can control. Should the resource ever move, the mapping will have to change behind the scenes, but her dependency on the Curated URI will remain stable.

3.3.3   EXAMPLE

The *Persistent URL* (PURL) system has been maintained by the *Online Computer Library Center* (OCLC)[7] for many years as a means of bridging the worlds of resolvable and stable identifiers. It was originally a fork of Apache 1.0, but was re-written in 2008 around the NetKernel[8] resource-oriented microkernel to modernize the architecture and add new features.

If you go to `http://purl.org`, you can create a user account and register PURLs within public or private domains. Domains represent a partition of the HTTP context. One popular public domain is `/net` which is often used to represent proxies for individuals on the Web.

For instance, `http://purl.org/net/bsletten` redirects via a `303 See Also` response to a document about me. Because the response is a 303 instead of a 200, HTTP clients can understand that the PURL was a valid reference to a non-network-addressable resource (i.e., me) and not to a document about me. The 303 takes the client to a document where they can learn more about the originally referenced resource (i.e., again, me). This currently returns a *Friend of a Friend* (FOAF)[9] profile for me. This is an RDF description of who I am, where I work, people I know, where I went to school, topics of interest, etc. While the redirection currently points to `http://bosatsu.net/foaf/brian.rdf`, in the future it may not. Anyone with a dependency upon the PURL will be redirected to any future location should they attempt to resolve it in the future.

The PURL system allows batch interfaces for bulk loading of PURL definitions. It is fairly widely used to ground RDF vocabulary terms in stable identifiers that can be redirected to a specific, resolvable vocabulary. A major use is by the Dublin Core Terms[10] such as `http://purl.org/dc/terms/creator`. The *National Center for Biological Ontology* (NCBO) maintains its own PURL

[7]`http://www.oclc.org`
[8]`http://netkernel.org`
[9]`http://foaf-project.org`
[10]`http://dublincore.org/documents/dcmi-terms/`

server instance at `http://purl.bioontology.org`, where it manages concept identifiers such as `http://purl.bioontology.org/ontology/MSH/D055963`.

3.3.4 CONSEQUENCES

The obvious consequence of the *Curated URI* pattern is constant redirection from the stable identifier to the current location. Resources that are accessed regularly in this manner may be painfully slow to use. A direct dependency may make more sense. Longer-term stability for less frequently accessed resources is a more natural fit.

    If you do find yourself running a curated system such as this and you need to avoid the constant redirection, it is possible to encourage caching on the client side of redirects. RFC2616[11] specifies in section 13.4 "Response Cacheability" that `Expires` and `Cache-Control` headers can be used for this purpose. 301s, 302s and 303s with future expiration dates can be cached. The support in browsers is fairly uneven, but it seems to be improving in newer browsers.[12] At best, browsers will not have to refollow the redirects always. At worst, they will have to.

3.4 INFORMED ROUTER

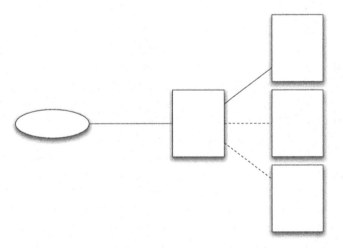

Figure 3.3: Informed Router Pattern.

---

[11]`http://tools.ietf.org/html/rfc2616`
[12]Steve Souders has a useful summary of the state of HTTP redirect caching here: `http://stevesouders.com/tests/redirects/results.php`.

3.4.1   INTENT

The *Informed Router* Pattern is similar to the function provided by a load balancer in conventional Web architectures. The difference is that the *Informed Router* is able to act upon domain-specific information not usually available to a generic routing engine.

3.4.2   MOTIVATION

The *Uniform Interface* of HTTP and the REST architectural style define a series of interaction patterns that apply across arbitrary resources. On the Web, it does not matter what kind of resource you are dealing with, browsers and similar clients can interact with financial reports, online games, social networking sites, and more in largely the same manner. Missing resources, authentication challenges, redirections, etc. can all be handled below the application level which allows the entire ecosystem to perform more consistently.

This predictability and the well-understood semantics of the HTTP methods allow reusable intermediaries to be developed. Enterprise caching proxies and load balancers are simple examples of the kind of value-added interception that can result from these architectural patterns. Caching usually requires the guidance of the origin servers to determine viability of the cache. Load balancers, however, can have more freedom of action. They might implement arbitrary routing algorithms depending upon their own optimization goals. A round robin [13] approach is a simple way to distribute stateless requests evenly across arbitrary backend resources by selecting each one in succession. If a server were to crash, it can be removed from the pool of options until it is fixed or rebooted.

While this is a simple and efficient approach to apply requests easily, it relies on them being largely the same basic burden on the handling resource. If some requests take longer than others or there is some other asymmetry to the infrastructure, this may not be the most efficient use of available servers. Another consequence of round robin scheduling is that it can be an inefficient use of memory for caching. If the same request is sent to each successive backend processor, we may use caching resources in each server to cache the same results. It is entirely possible to route requests based upon the source IP address, the destination IP address, the requested resource, or some other kind of token the server can send to the client to return on subsequent requests. A more insightful load balancer could route based upon server load, average response time, or to encourage caching by returning clients to the same server each time they return.

Another aspect of the REST architectural style is the requirement for self-describing messages to be sent back and forth to represent the state of the resource. This constraint is an important form of decoupling that also extends the options for intermediaries to manipulate, handle or, in the case of this pattern, direct a response based on the payload of the message.

A self-describing message is one that contains all of the resource state necessary to handle the request. There is no out-of-band sharing of information that requires special knowledge that intermediaries (as opposed to the client and server) would not also be privy to. The self-description

---

[13]http://en.wikipedia.org/wiki/Round-robin_scheduling

will include the headers in the request such as an identifiable Content-Type value that identifies the nature of the representation.

By being able to identify the type of message and the context in which it is being sent or received, an intermediary could easily annotate, filter, or manipulate the message as it passes through in either direction. To do anything meaningful with a non-standard type, the intermediary will need to have custom handling for the type, but generic services such as spell-checking, scanning for sensitive information, regulatory compliance, regular expression matching, etc. are possible across arbitrary outgoing text-oriented messages.

The ability to introspect on the type of the message being submitted to a server allows for content-based routing. This is generally not something people think about in the context of Web architectures, even though it is more widely used in more traditionally messaging environments. It is entirely possible, however, and even designed into the interaction patterns.

3.4.3   EXAMPLE

An example of the *Informed Router* Pattern might be to route requests based upon arbitrary prioritization schemes. These will be based upon domain-specific criteria that conventional load-balancing and routing intermediaries would be unable to use. In the case of an e-commerce site, a customer's status or the vendor of the requested product might be used as triggers to route the message to a higher-priority handler. Whether this was done by explicitly sending the request to a different physical endpoint or through the annotation of the message with special metadata does not really matter. This kind of interception pattern has historically been available to programmatic endpoints like the Java Servlet Filter model,[14] but that is bound very tightly to a particular implementation technology. We would like to consider implementing the routing strategy at a separate level. Because we are operating directly on the message, these architectural solutions can be uncoupled from the currently used technology stack for longer-term reuse.

When combined with the *Resource Heatmap* Pattern, dynamic routing can be applied to "hot" resources to offload unexpected imbalances in the patterns of requested content. Imagine a generally round-robin load-balancing policy that automatically detected spikes in traffic to particular resources. Otherwise, undifferentiated handlers could be partitioned to force the hotter resources to a specific collection of boxes to encourage caching and rebalance the request load activity. This could be done dynamically and only for the lifetime of the surge. If the infrastructure used cloud-based resources, it could spin up new instances and route to them for the specified resource. When the demand slowed down, these temporary cloud-based resources could be spun back down to reduce cost so that it more closely tracked actual demand.

Beyond handling spikes in activity and server resources, we can imagine routing requests to handlers in response to context-specific fluctuations. This might include content and concept-based routing of domain-specific entities to established subject-matter experts. That expertise mapping

---

[14]http://en.wikipedia.org/wiki/Java_Servlet

might be prioritized by organizational discovery of influence,[15] but updated in consideration of employee vacation and travel schedules.

As may be obvious, we are attempting to have the best of both worlds. On the one hand, we want the standardization and predictability of the Web's *Uniform Interface*. On the other, we would like to take advantage of domain-specific nuances to improve our capacity to respond to changing environments and contexts. The reusable infrastructure can be tweaked by *Informed Router* Pattern implementations. Even if all of the components ultimately react externally with the same, predictable behavior, internally, we can re-route, re-interpret, and re-prioritize how we handle requests by considering the detail of a specifics available to us.

3.4.4   CONSEQUENCES

As with any dynamic system, it is important to build tolerance ranges into the shifts in routing to avoid unnecessary churn. Rather than modifying the policies when some simple threshold is crossed, you will want to factor in historical trends, the duration of increased activity, etc. to achieve smooth, meaningful transitions rather than spasmodically variant, noisy activity.

From a security perspective, privileging how messages are handled based on information received from the client is a risky endeavor. Without some level of trust in the authenticity of the request, clients could provide false inputs or intermediaries could modify the request maliciously. In such a scenario, a client might achieve privilege escalation or a rogue intermediary could violate the integrity of the initial message. Most Web-based solutions enforce authentication and authorization at the transport level. If this is a concern, you might require that clients digitally sign the body of their requests. Intermediaries may manipulate messages in verifiable ways as well by digitally signing the modified portions for auditing purposes.

See also: *Resource Heatmap* (2.4)

3.5   WORKFLOW

3.5.1   INTENT

The *Workflow* Pattern encodes a series of steps into a resource abstraction where the client learns what options are available through the resource representation. The server is in charge of enabling and disabling state transitions based on the context of the requests, the client choices, and other inputs.

3.5.2   MOTIVATION

We wrap up by looking at a pattern based upon the fundamentals of the Web beyond URLs and HTTP. We consider not just how we identify and send representations around, but also, how do we discover what options are available to us. It is a little overwhelming to consider how many different ways people use standard browsers on the Web, either individually or in collaborative interactions.

---

[15] See the discussion in the *Resource Heatmap* Pattern (2.4).

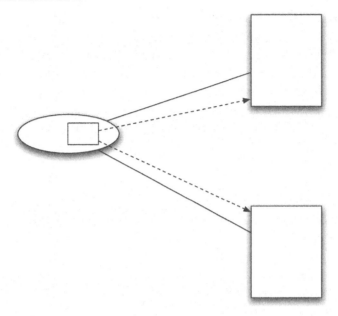

Figure 3.4: Workflow Pattern.

The thing that drives all of this is *Hypermedia*. This is a fancy term to describe characteristics of the representation format and the interaction styles that are discovered within. In the *Information Resource* Pattern (1.2) section, we saw the term *Hypermedia Factor* (HFactor) to describe specific types of interactions found in hypermedia formats. This term comes from Mike Amundsen in Amundsen [2011] who details nine specific HFactors shown in Figure 3.5. These factors give browsers the ability to discover the *affordances*, or user-interface choices and capabilities to provide to a user.

While we are not going to exhaustively walk through each of the following,

1. **Embedded Links (LE)**

2. **Outbound Links (LO)**

3. **Templated Links (LT)**

4. **Non-Idempotent Links (LN)**

5. **Idempotent Links (LI)**

6. **Read Controls (CR)**

7. **Update Controls (CU)**

8. **Method Controls (CM)**

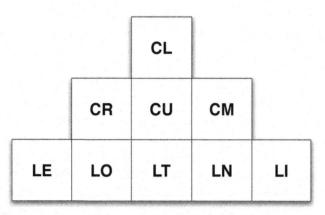

Figure 3.5: Hypermedia Factors (HFactors) in various representations.

the individual factors contribute to the expressive powers of a particular format that supports them. For example, HTML support looks like Figure 3.6.

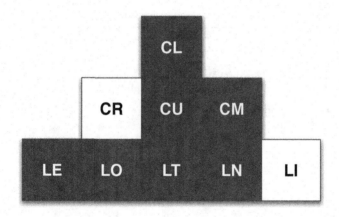

Figure 3.6: Hypermedia Factors (HFactors) in HTML.

It has the ability to specify embedded links (e.g., images, stylesheets), outbound links, and templated links with various controls (e.g., forms). It does not allow the use of the DELETE or PUT methods,[16] so it lacks support for Idempotent Links (LI). It also does not allow you to do content negotiation on GET requests, so it lacks support for Read Controls (CR). Otherwise, it is, unsurprisingly, a very well-featured hypermedia format.[17]

---

[16]There was interest in adding support for these methods in HTML5, but editorial whim kept them out.
[17]You are encouraged to consult Amundsen [2011] for a deeper discussion of hypermedia systems.

The factors present in HTML combine to give us the ability to specify all of the photo sharing, chatting, headline reading, banking, gaming, and everything else we do on the Web. If an application requires the user to log in, that is all that is presented to them. Once they log in, there may be additional capabilities unlocked. Entire workflows are expressed based upon the presence or absence of certain elements and links.

A proper REST API is driven by similar mechanisms. A hypermedia format presents the client with affordances to offer to the user. Links are generally annotated with rel attributes that advertise the intended use through an established protocol. The links themselves are discovered in the bodies of the representations. Clients are not aware of how to build them up. This allows the server to rearrange where resources are located without negatively affecting the client, just like a website can move its images around without affecting browsers. The mechanism promotes loose coupling between the clients and servers. We do have coupling in these kinds of systems, they are just to the media types, not any particular server layout. The considerable effort that goes into a browser to support HTML is what ultimately makes the applications work.

As we discussed in the *Information Resource* Pattern (1.2), there are no standard, general-purpose, hypermedia formats at the moment, although there are some widely used proposals. Mike Kelly[18] has designed a hypermedia format for REST APIs called the *Hypertext Application Language* (HAL).[19]

Here is an example of a set of orders expressed in the JSON form of HAL (application/hal+json):

```
 1  {
 2    "_links": {
 3      "self": { "href": "/orders" },
 4      "next": { "href": "/orders?page=2" },
 5      "find": { "href": "/orders{?id}", "templated": true },
 6      "admin": [
 7        { "href": "/admins/2", "title": "Fred" },
 8        { "href": "/admins/5", "title": "Kate" }
 9      ]
10    },
11    currentlyProcessing: 14,
12    shippedToday: 20,
13    "_embedded": {
14    "orders": [{
15        "_links": {
16          "self": { "href": "/orders/123" },
17          "basket": { "href": "/baskets/98712" },
18          "customer": { "href": "/customers/7809" }
19        },
20        "total": 30.00,
21        "currency": "USD",
22        "status": "shipped",
23      },{
24        "_links": {
25          "self": { "href": "/orders/124" },
26          "basket": { "href": "/baskets/97213" },
```

[18]http://stateless.co
[19]http://stateless.co/hal_specification.html

```
27          "customer": { "href": "/customers/12369" }
28        },
29        "total": 20.00,
30        "currency": "USD",
31        "status": "processing"
32      }]
33    }
34  }
```

You will recognize many of the ideas we described for the application.vnd.collections+json format in the *Collections Resource* Pattern (1.3). We have support for link discovery, pagination, related resources, arbitrary key value pairs, etc.

Kevin Swiber[20] has developed a different representation format called SIREN.[21] A JSON form of SIREN (application/vnd.siren+json) representing an order might look like the following:

```
1   {
2     "class": [ "order" ],
3     "properties": {
4         "orderNumber": 42,
5         "itemCount": 3,
6         "status": "pending"
7     },
8     "entities": [
9       {
10        "class": [ "items", "collection" ],
11        "rel": [ "http://x.io/rels/order-items" ],
12        "href": "http://api.x.io/orders/42/items"
13      },
14      {
15        "class": [ "info", "customer" ],
16        "rel": [ "http://x.io/rels/customer" ],
17        "properties": {
18          "customerId": "pj123",
19          "name": "Peter Joseph"
20        },
21        "links": [
22          { "rel": [ "self" ], "href": "http://api.x.io/customers/pj123" }
23        ]
24      }
25    ],
26    "actions": [
27      {
28        "name": "add-item",
29        "title": "Add Item",
30        "method": "POST",
31        "href": "http://api.x.io/orders/42/items",
32        "type": "application/x-www-form-urlencoded",
33        "fields": [
34          { "name": "orderNumber", "type": "hidden", "value": "42" },
35          { "name": "productCode", "type": "text" },
36          { "name": "quantity", "type": "number" }
```

---

[20]https://github.com/kevinswiber/
[21]https://github.com/kevinswiber/siren

```
37          ]
38        }
39      ],
40      "links": [
41        { "rel": [ "self" ], "href": "http://api.x.io/orders/42" },
42        { "rel": [ "previous" ], "href": "http://api.x.io/orders/41" },
43        { "rel": [ "next" ], "href": "http://api.x.io/orders/43" }
44      ]
45    }
```

We again see support for collections, pagination, and related links, but we also see explicit support for arbitrary actions. In this case, to add an item to the order, you would submit the specified fields as a URL-encoded form submission to the HREF specified via a POST method.

The thing that is easy to miss is that these formats are useful across domains and workflows. There is nothing implementation-specific about them. With their hypermedia links and self-describing mechanisms, they allow a client to learn what options are available. They represent a server-driven way of navigating changing states on the resource. This is an implementation of the *Workflow* Pattern. It is effectively simply a hypermedia-based system, but it is important enough on its own right to deserve its own pattern.

Recall that the resource abstraction is a superset of data, documents, services, and concepts. We are able to interact with all of these resource types using semantically meaningful, constrained, predictable methods. The links in a hypermedia workflow can encompass all of this variance, which helps us to orchestrate workflows across disparate backend systems. By using some of the other patterns in this book, we can transform, organize, and protect systems that know nothing about these uses.

The expression of a workflow is not necessarily the prescription of a user interface. While we may wish to allow or disallow certain actions based upon resource state or business rules, we do not want to mandate how a user actually interacts with their client. We are defining a "mechanism not a policy."[22] We want the workflow to allow alternate paths through the transitions, but only along valid lines.

3.5.3   EXAMPLE

As an example, we will consider a commerce engine for a generic retail system. While you are encouraged to use HAL, SIREN, Collections+JSON, and other reusable representations, for simplicity, we will imagine a custom representation design. Most of the ideas we discuss will transcend specific representation and could be easily mapped to one of these other formats.

A commerce system must connect consumers to products. We need to allow them to place orders, we want to alert them to promotions and bundles. We want to make it easy to track past orders. While we are not going to design the entire system here, these are the kinds of things we must consider.

---

[22]http://c2.com/cgi/wiki?MechanismNotPolicy

Like most websites, we want a well-known, stable starting point. From there, clients will discover what they are allowed to do and the transitions they can make. Our starting point will be: http://example.com/commerce.[23] The returned representation might look something like:

```
1  <commerce version="1.0">
2     <link rel="login" href="http://example.com/login"/>
3     <link rel="order" href="http://example.com/order/"/>
4  </commerce>
```

Our initial representation gives the user two options. They are able to log in or create a new order. A GET request to the "login" link could trigger a 401 response with support for *HTTP Basic Auth, HTTP Digest, Two-legged OAuth*, etc. But, because we want to provide mechanisms without specifying a policy, the client application (as a proxy for the user) can decide whether they wish to log into the system or create a guest checkout order.

We define a protocol around the representation that specifies a POST of a complete order to the order link is acceptable, as is a POST of an empty order. Either one will trigger the presence of a current order. Subsequent GET requests to the starting point will yield a new link:

```
1  <commerce version="1.0">
2     <link rel="login" href="http://example.com/login"/>
3     <link rel="order" href="http://example.com/order/"/>
4     <link rel="current" href="http://example.com/order/id/12345-6789"/>
5  </commerce>
```

If the user logged in before creating the order, we can associate the order with the account. In the future, a subsequent login request can resupply the same link, however we will probably not show the current link until they log in. If the user chooses a guest checkout workflow, closes the application, and comes back in the future, a cookie could be used to re-associate the anonymous order with them. Or, we could just trash unfulfilled guest orders after a few hours. It all depends upon our intended user experience, security concerns, etc.[24] If the user creates a guest order, adds some items to her cart, and then logs in, we will want to transition the guest order to the user's account. All of these options are possible with what we have defined so far.

Keep in mind this is a cartoon of such a system, as we do not actually have the time or space to fully design such a system. But, we have an established MIME type (e.g., application/vnd.example.commerce+XML) that can be documented. Clients can issue standard GET requests to the starting point and react to receiving one of these. They will know to look for these links with certain rel values. This initial interaction can also be extended by supporting content negotiation and adding JSON, HAL, SIREN, or whatever other format you want. This is one of the reasons why I do not have a problem with custom MIME types. As standards become adopted, it is trivial to add support for them without breaking existing clients. They can decide to move to adopt the standard as they wish.

---

[23]We ignore the bad habit of versioning APIs. Webpages are not generally versioned... because they do not need to be.
[24]Don't forget that creating database records from unauthenticated clients might open you up for denial of service (DOS) attacks!

What about our lack of versioning in the API URLs? Well, notice we can still express this as metadata on the response to an attribute, but ideally, we would like to ignore versioning to the extent that we can. What if we want to add some new capabilities to the commerce representation? One of the guidelines of Web development is that clients should ignore things they are not expecting. Clients that honor this will not mind if we start sending back an extra link:

```
1  <commerce version="1.0">
2     <link rel="login" href="http://example.com/login"/>
3     <link rel="order" href="http://example.com/order/"/>
4     <link rel="current" href="http://example.com/order/id/12345-6789"/>
5     <link rel="search" href="http://example.com/search{;keyword}"/>
6  </commerce>
```

Because we are not using a representation like HAL or SIREN, deployed clients will not know how to respond to the presence of a new link like this. However, they should not break either. They simply ignore it. Clients that wish to roll out support at the user interface level for the new search capability can do so when and how they wish. Their release is not coupled to our release.

We have adopted an RFC6570[25]-complaint URI template. Clients that understand this can collect the keywords however they like and then substitute them into the URI template before issuing a GET request. This gives us permission to move the URLs around in the future without breaking clients, as long as we keep the attributes we are looking for the same. In the new location, the client will still simply substitute the collected values in a new location.

If the user logs in, she could be presented with a link to her account:

```
1  <commerce version="1.0">
2     <link rel="account" href="http://example.com/account/id/22101"/>
3     <link rel="order" href="http://example.com/order/"/>
4     <link rel="current" href="http://example.com/order/id/12345-6789"/>
5     <link rel="search" href="http://example.com/search{;keyword}"/>
6  </commerce>
```

This information might come back as an application.vnd.example.account+XML representation, JSON or anything else. Each of these links is its own resource and can have its own content negotiation policies. Regardless, we discover links to the current order, recent orders, all orders, promotions on the account, etc.

```
1   <account version="1.0">
2      <link rel="self" href="http://example.com/account/id/22101"/>
3      <username>bsmith</username>
4      <contact>
5       <email>bsmith@someplace.com</email>
6       <twitter>bevsmithnotexist</twitter>
7      </contact>
8      <orders>
9         <link rel="all" href="http://example.com/order/account/id/22101"/>
10        <link rel="recent"
              href="http://example.com/order/account/id/22101;recent=6mo"/>
```

[25]http://tools.ietf.org/html/rfc6570

```
11          <order  type="current"  id="12345−6789"
                href="http://example.com/order/id/12345−6789">
12              ...
13          </order>
14
15      </orders>
16      <promotions  href="http://example.com/promotion/account/id/22101"/>
17  </account>
```

If this user has a problem with an order and she calls into a customer service center, the agent will use an application that takes a reference to the user account. It fetches the information and provides the details of the current order as well as a hook to find others. If during the course of the conversation the agent decides to award the customer a promotion, his application discovers the link that will accept promotions to associate with the account. That is a privileged action, however. We do not want this user's friends who work for our retailer being able to give her discounts she is not due. Just because we provide a link does not mean it can be freely used. Issuing the POST will require a certain type of credential indicating the agent's role within the organization.

We could obviously extend this example further, but the point is that we have a resource-driven workflow that makes sense beyond any particular implementation technology choices. We are free to migrate backend systems as long as we update the resource handler to use them. Clients can, hopefully, remain blessedly ignorant of such changes.

As we have belabored above, ideally, we would be using a standard, hypermedia-driven representation, but even if we do not, we gain the benefit of a large percentage of the Web infrastructure and all of the properties it entails. We have defined a mechanism that can survive for quite some time in our organization. The expression of workflow can change to reflect new business contexts, policies, opportunities, etc., but the clients can be written in ways that make them resilient to such changes (or able to adopt them on their own terms).

It is not an accident that the Web has been successful at a level of diversity and scale that would have been impossible to predict. By accumulating the various design choices into our own systems, we can imbue them with similar properties. It seems foolish to do otherwise in the 21st century.

3.5.4   CONSEQUENCES

The *Workflow* Pattern takes considerable thinking to abstract away from implementation details and to imagine what might be useful beyond an application's immediate needs. The cost is generally front-loaded but amortized across the lifetime of the resources which, if done correctly, should be considerably longer than typical information technology resources, services, etc.

The major contrast to this style embraced by traditional software developers is to create application-specific, point-to-point interactions. Information and content sources are integrated at the database or object model layers and then projected to the client in a specific, prescribed format and schema. They are scoped to the immediate needs of the application, clients, etc. If they need to change something in the future, they will simply refactor all relevant components. The initial cost may be lower, but you are constantly refactoring and potentially disrupting clients who may have completely different release cycles[26].

The Web is a living reference implementation of a platform for loosely-coupled, scalable, independently-deployed components. Not everything fits neatly in coarse-grained, synchronous, client-server interactions, but the patterns in this book should have helped you realize that is not an entirely fair critique of Web-based systems anymore. It all comes down to the fact that our systems are projections of our needs from current business and technological contexts. There is no question that these will change over time. The Web itself has spurred technical development to manage both its successes (e.g., TCP/IP traffic congestion, slow start and other protocol extensions to meet scalability demands) as well as its limitations (e.g., long-lived sockets, Web Sockets, HTTP 2.0 to deal with client-initiated limitations).

While it may take an effort, embracing a suite of technologies that embrace change should continue to pay dividends over time. Our systems should be more flexible, more extensible and more amenable to the crazy and unpredictable shifts in interest for the resources we produce and consume. The *Workflow* Pattern reflects these ideas and should provide many of the same benefits we see on the public Web in our own private Webs.

---

[26]Or would at least prefer to!

Bibliography

T. Berners-Lee, "Information Management : A Proposal," http://www.w3.org/History/1989/proposal.html. xvi

G. Hohpe, B. Woolf, *Enterprise Integration Patterns: Designing, Building, and Deploying Messaging Solutions*, Boston, MA: Addison-Wesley Longman Publishing Co., Inc., 2003.

J. Webber, S. Parastatidis, I. Robinson, *REST In Practice : Hypermedia and Systems Architecture*, Sebastapol, CA: O'Reilly Media, Inc., 2010.

T. Heath, C. Bizer, *Linked Data: Evolving the Web into a Global Data Space*, Morgan & Claypool, 2011. DOI: 10.2200/S00334ED1V01Y201102WBE001 18

M. Amundsen, *Hypermedia APIs with HTML5 & Node*, Sebastapol, CA: O'Reilly Media, Inc., 2011. 6, 7, 10, 64, 65

L. Richardson, M. Amundsen, *RESTful Web APIs*, Sebastapol, CA: O'Reilly Media, Inc., 2013. 6

S. Allamaraju, *RESTful Web Services Cookbook: Solutions for Improving Scalability and Simplicity*, Sebastapol, CA: O'Reilly Media, Inc., 2010. 7

Author's Biography

BRIAN SLETTEN

Brian Sletten is a liberal arts-educated software engineer with a focus on using and evangelizing forward-leaning technologies. He has a background as a system architect, a developer, a security consultant, a mentor, a team lead, an author, and a trainer and operates in all of those roles as needed. His experience has spanned the online game, defense, finance, academic, hospitality, retail, and commercial domains. He has worked with a wide variety of technologies such as network matrix switch controls, 3D simulation/visualization, Grid Computing, P2P, and Semantic Web-based systems. He has a B.S. in Computer Science from the College of William and Mary. He is President of Bosatsu Consulting, Inc. and lives in Auburn, CA.

He focuses on web architecture, resource-oriented computing, social networking, the Semantic Web, scalable systems, security consulting, and other technologies of the late 20th and early 21st centuries.